BRITANNICA
Mathematics in Context

Packages and Polygons

⊕ Britannica
ENCYCLOPÆDIA BRITANNICA EDUCATIONAL CORPORATION

Mathematics in Context is a comprehensive curriculum for the middle grades. It was developed in collaboration with the Wisconsin Center for Education Research, School of Education, University of Wisconsin–Madison and the Freudenthal Institute at the University of Utrecht, The Netherlands, with the support of National Science Foundation Grant No. 9054928.

National Science Foundation

Opinions expressed are those of the authors
and not necessarily those of the Foundation

ISBN 0-7826-1533-3
1 2 3 4 5 WK 02 01 00 99 98

The *Mathematics in Context* Development Team

Mathematics in Context is a comprehensive curriculum for the middle grades. The National Science Foundation funded the National Center for Research in Mathematical Sciences Education at the University of Wisconsin–Madison to develop and field-test the materials from 1991 through 1996. The Freudenthal Institute at the University of Utrecht in The Netherlands, as a subcontractor, collaborated with the University of Wisconsin–Madison on the development of the curriculum.

The initial version of *Packages and Polygons* was developed by Martin Kindt. It was adapted for use in American schools by Mary S. Spence, Laura J. Brinker, and Gail Burrill.

National Center for Research in Mathematical Sciences Education Staff

Thomas A. Romberg
Director

Joan Daniels Pedro
Assistant to the Director

Gail Burrill
Coordinator
Field Test Materials

Margaret R. Meyer
Coordinator
Pilot Test Materials

Mary Ann Fix
Editorial Coordinator

Sherian Foster
Editorial Coordinator

James A. Middleton
Pilot Test Coordinator

Margaret A. Pligge
First Edition Coordinator

Project Staff

Jonathan Brendefur
Laura J. Brinker
James Browne
Jack Burrill
Rose Byrd
Peter Christiansen
Barbara Clarke
Doug Clarke
Beth R. Cole

Fae Dremock
Jasmina Milinkovic
Kay Schultz
Mary C. Shafer
Julia A. Shew
Aaron N. Simon
Marvin Smith
Stephanie Z. Smith
Mary S. Spence
Kathleen A. Steele

Freudenthal Institute Staff

Jan de Lange
Director

Els Feijs
Coordinator

Martin van Reeuwijk
Coordinator

Project Staff

Mieke Abels
Nina Boswinkel
Frans van Galen
Koeno Gravemeijer
Marja van den Heuvel-Panhuizen
Jan Auke de Jong
Vincent Jonker
Ronald Keijzer

Martin Kindt
Jansie Niehaus
Nanda Querelle
Anton Roodhardt
Leen Streefland
Adri Treffers
Monica Wijers
Astrid de Wild

Acknowledgments

Several school districts used and evaluated one or more versions of the materials: Ames Community School District, Ames, Iowa; Parkway School District, Chesterfield, Missouri; Stoughton Area School District, Stoughton, Wisconsin; Madison Metropolitan School District, Madison, Wisconsin; Milwaukee Public Schools, Milwaukee, Wisconsin; and Dodgeville School District, Dodgeville, Wisconsin. Two sites were involved in staff developments as well as formative evaluation of materials: Culver City, California, and Memphis, Tennessee. Two sites were developed through partnership with Encyclopædia Britannica Educational Corporation: Miami, Florida, and Puerto Rico. University Partnerships were developed with mathematics educators who worked with preservice teachers to familiarize them with the curriculum and to obtain their advice on the curriculum materials. The materials were also used at several other schools throughout the United States.

We at Encyclopædia Britannica Educational Corporation extend our thanks to all who had a part in making this program a success. Some of the participants instrumental in the program's development are as follows:

Allapattah Middle School
Miami, Florida
Nemtalla (Nikolai) Barakat

Ames Middle School
Ames, Iowa
Kathleen Coe
Judd Freeman
Gary W. Schnieder
Ronald H. Stromen
Lyn Terrill

Bellerive Elementary
Creve Coeur, Missouri
Judy Hetterscheidt
Donna Lohman
Gary Alan Nunn
Jakke Tchang

Brookline Public Schools
Brookline, Massachusetts
Rhonda K. Weinstein
Deborah Winkler

Cass Middle School
Milwaukee, Wisconsin
Tami Molenda
Kyle F. Witty

Central Middle School
Waukesha, Wisconsin
Nancy Reese

Craigmont Middle School
Memphis, Tennessee
Sharon G. Ritz
Mardest K. VanHooks

Crestwood Elementary
Madison, Wisconsin
Diane Hein
John Kalson

Culver City Middle School
Culver City, California
Marilyn Culbertson
Joel Evans
Joy Ellen Kitzmiller
Patricia R. O'Connor
Myrna Ann Perks, Ph.D.
David H. Sanchez
John Tobias
Kelley Wilcox

Cutler Ridge Middle School
Miami, Florida
Lorraine A. Valladares

Dodgeville Middle School
Dodgeville, Wisconsin
Jacqueline A. Kamps
Carol Wolf

Edwards Elementary
Ames, Iowa
Diana Schmidt

Fox Prairie Elementary
Stoughton, Wisconsin
Tony Hjelle

Grahamwood Elementary
Memphis, Tennessee
M. Lynn McGoff
Alberta Sullivan

Henry M. Flagler Elementary
Miami, Florida
Frances R. Harmon

Horning Middle School
Waukesha, Wisconsin
Connie J. Marose
Thomas F. Clark

Huegel Elementary
Madison, Wisconsin
Nancy Brill
Teri Hedges
Carol Murphy

Hutchison Middle School
Memphis, Tennessee
Maria M. Burke
Vicki Fisher
Nancy D. Robinson

Idlewild Elementary
Memphis, Tennessee
Linda Eller

Jefferson Elementary
Santa Ana, California
Lydia Romero-Cruz

Jefferson Middle School
Madison, Wisconsin
Jane A. Beebe
Catherine Buege
Linda Grimmer
John Grueneberg
Nancy Howard
Annette Porter
Stephen H. Sprague
Dan Takkunen
Michael J. Vena

Jesus Sanabria Cruz School
Yabucoa, Puerto Rico
Andreíta Santiago Serrano

John Muir Elementary School
Madison, Wisconsin
Julie D'Onofrio
Jane M. Allen-Jauch
Kent Wells

Kegonsa Elementary
Stoughton, Wisconsin
Mary Buchholz
Louisa Havlik
Joan Olsen
Dominic Weisse

Linwood Howe Elementary
Culver City, California
Sandra Checel
Ellen Thireos

Mitchell Elementary
Ames, Iowa
Henry Gray
Matt Ludwig

New School of Northern Virginia
Fairfax, Virginia
Denise Jones

Northwood Elementary
Ames, Iowa
Eleanor M. Thomas

Orchard Ridge Elementary
Madison, Wisconsin
Mary Paquette
Carrie Valentine

Parkway West Middle School
Chesterfield, Missouri
Elissa Aiken
Ann Brenner
Gail R. Smith

Ridgeway Elementary
Ridgeway, Wisconsin
Lois Powell
Florence M. Wasley

Roosevelt Elementary
Ames, Iowa
Linda A. Carver

Roosevelt Middle
Milwaukee, Wisconsin
Sandra Simmons

Ross Elementary
Creve Coeur, Missouri
Annette Isselhard
Sheldon B. Korklan
Victoria Linn
Kathy Stamer

St. Joseph's School
Dodgeville, Wisconsin
Rita Van Dyck
Sharon Wimer

St. Maarten Academy
St. Peters, St. Maarten, NA
Shareed Hussain

Sarah Scott Middle School
Milwaukee, Wisconsin
Kevin Haddon

Sawyer Elementary
Ames, Iowa
Karen Bush Hoiberg

Sennett Middle School
Madison, Wisconsin
Brenda Abitz
Lois Bell
Shawn M. Jacobs

Sholes Middle School
Milwaukee, Wisconsin
Chris Gardner
Ken Haddon

Stephens Elementary
Madison, Wisconsin
Katherine Hogan
Shirley M. Steinbach
Kathleen H. Vegter

Stoughton Middle School
Stoughton, Wisconsin
Sally Bertelson
Polly Goepfert
Jacqueline M. Harris
Penny Vodak

Toki Middle School
Madison, Wisconsin
Gail J. Anderson
Vicky Grice
Mary M. Ihlenfeldt
Steve Jernegan
Jim Leidel
Theresa Loehr
Maryann Stephenson
Barbara Takkunen
Carol Welsch

Trowbridge Elementary
Milwaukee, Wisconsin
Jacqueline A. Nowak

W. R. Thomas Middle School
Miami, Florida
Michael Paloger

Wooddale Elementary Middle School
Memphis, Tennessee
Velma Quinn Hodges
Jacqueline Marie Hunt

Yahara Elementary
Stoughton, Wisconsin
Mary Bennett
Kevin Wright

Site Coordinators

Mary L. Delagardelle—Ames Community Schools, Ames, Iowa

Dr. Hector Hirigoyen—Miami, Florida

Audrey Jackson—Parkway School District, Chesterfield, Missouri

Jorge M. López—Puerto Rico

Susan Militello—Memphis, Tennessee

Carol Pudlin—Culver City, California

Reviewers and Consultants

Michael N. Bleicher
Professor of Mathematics
University of Wisconsin–Madison
Madison, WI

Diane J. Briars
Mathematics Specialist
Pittsburgh Public Schools
Pittsburgh, PA

Donald Chambers
Director of Dissemination
University of Wisconsin–Madison
Madison, WI

Don W. Collins
Assistant Professor of Mathematics Education
Western Kentucky University
Bowling Green, KY

Joan Elder
Mathematics Consultant
Los Angeles Unified School District
Los Angeles, CA

Elizabeth Fennema
Professor of Curriculum and Instruction
University of Wisconsin–Madison
Madison, WI

Nancy N. Gates
University of Memphis
Memphis, TN

Jane Donnelly Gawronski
Superintendent
Escondido Union High School
Escondido, CA

M. Elizabeth Graue
Assistant Professor of Curriculum and Instruction
University of Wisconsin–Madison
Madison, WI

Jodean E. Grunow
Consultant
Wisconsin Department of Public Instruction
Madison, WI

John G. Harvey
Professor of Mathematics and Curriculum & Instruction
University of Wisconsin–Madison
Madison, WI

Simon Hellerstein
Professor of Mathematics
University of Wisconsin–Madison
Madison, WI

Elaine J. Hutchinson
Senior Lecturer
University of Wisconsin–Stevens Point
Stevens Point, WI

Richard A. Johnson
Professor of Statistics
University of Wisconsin–Madison
Madison, WI

James J. Kaput
Professor of Mathematics
University of Massachusetts–Dartmouth
Dartmouth, MA

Richard Lehrer
Professor of Educational Psychology
University of Wisconsin–Madison
Madison, WI

Richard Lesh
Professor of Mathematics
University of Massachusetts–Dartmouth
Dartmouth, MA

Mary M. Lindquist
Callaway Professor of Mathematics Education
Columbus College
Columbus, GA

Baudilio (Bob) Mora
Coordinator of Mathematics & Instructional Technology
Carrollton-Farmers Branch Independent School District
Carrollton, TX

Paul Trafton
Professor of Mathematics
University of Northern Iowa
Cedar Falls, IA

Norman L. Webb
Research Scientist
University of Wisconsin–Madison
Madison, WI

Paul H. Williams
Professor of Plant Pathology
University of Wisconsin–Madison
Madison, WI

Linda Dager Wilson
Assistant Professor
University of Delaware
Newark, DE

Robert L. Wilson
Professor of Mathematics
University of Wisconsin–Madison
Madison, WI

TABLE OF CONTENTS

BRITANNICA
Mathematics
in
Context

Dear Teacher,

Welcome! *Mathematics in Context* is designed to reflect the National Council of Teachers of Mathematics Standards for School Mathematics and to ground mathematical content in a variety of real-world contexts. Rather than relying on you to explain and demonstrate generalized definitions, rules, or algorithms, students investigate questions directly related to a particular context and construct mathematical understanding and meaning from that context.

The curriculum encompasses 10 units per grade level. *Packages and Polygons* is designed to be the first in the geometry strand for grade 7/8, but the unit also lends itself to independent use to introduce students to identifying the properties of polygons and polyhedra, constructing geometric models, and drawing two- and three-dimensional figures.

In addition to the Teacher Guide and Student Books, *Mathematics in Context* offers the following components that will inform and support your teaching:

- *Teacher Resource and Implementation Guide,* which provides an overview of the complete system, including program implementation, philosophy, and rationale

- *Number Tools,* Volumes 1 and 2, which are a series of blackline masters that serve as review sheets or practice pages involving number issues/basic skills

- *News in Numbers,* which is a set of additional activities that can be inserted between or within other units; it includes a number of measurement problems that require estimation.

Thank you for choosing *Mathematics in Context.* We wish you success and inspiration!

Sincerely,

The Mathematics in Context Development Team

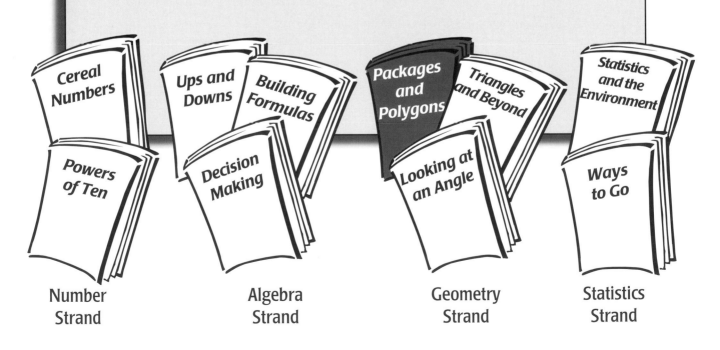

Number Strand Algebra Strand Geometry Strand Statistics Strand

Overview

B R I T A N N I C A

Mathematics in Context

How to Use This Book

This unit is one of 40 for the middle grades. Each unit can be used independently; however, the 40 units are designed to make up a complete, connected curriculum (10 units per grade level). There is a Student Book and a Teacher Guide for each unit.

Each Teacher Guide comprises elements that assist the teacher in the presentation of concepts and in understanding the general direction of the unit and the program as a whole. Becoming familiar with this structure will make using the units easier.

Each Teacher Guide consists of six basic parts:

- Overview
- Student Materials and Teaching Notes
- Assessment Activities and Solutions
- Glossary
- Blackline Masters
- Try This! Solutions

Overview

Before beginning this unit, read the Overview in order to understand the purpose of the unit and to develop strategies for facilitating instruction. The Overview provides helpful information about the unit's focus, pacing, goals, and assessment, as well as explanations about how the unit fits with the rest of the *Mathematics in Context* curriculum.

Student Materials and Teaching Notes

This Teacher Guide contains all of the student pages (except the Try This! activities), each of which faces a page of solutions, samples of students' work, and hints and comments about how to facilitate instruction. Note: Solutions for the Try This! activities can be found at the back of this Teacher Guide.

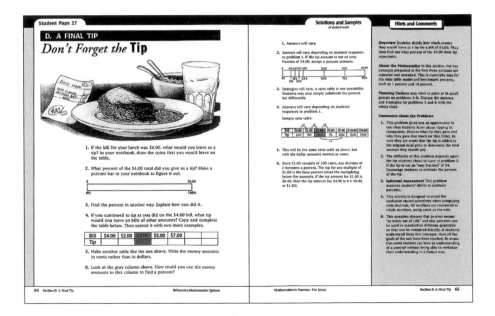

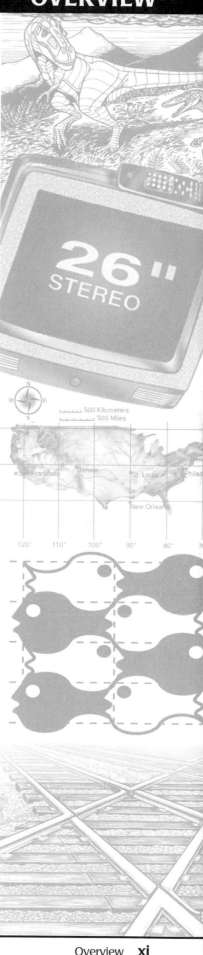

Each section within the unit begins with a two-page spread that describes the work students do, the goals of the section, new vocabulary, and materials needed, as well as providing information about the mathematics in the section and ideas for pacing, planning instruction, homework, and assessment.

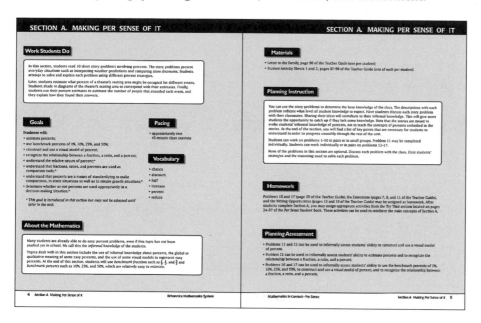

Assessment Activities and Solutions

Information about assessment can be found in several places in this Teacher Guide. General information about assessment is given in the Overview; informal assessment opportunities are identified on the teacher pages that face each student page; and the Assessment Activities section of this guide provides formal assessment opportunities.

Glossary

The Glossary defines all vocabulary words listed on the Section Opener pages. It includes the mathematical terms that may be new to students, as well as words associated with the contexts introduced in the unit. (Note: The Student Book does not have a glossary. This allows students to construct their own definitions, based on their personal experiences with the unit activities.)

Blackline Masters

At the back of this Teacher Guide are blackline masters for photocopying. The blackline masters include a letter to families (to be sent home with students before beginning the unit), several student activity sheets, and assessment masters.

Try This! Solutions

Also included in the back of this Teacher Guide are the solutions to several Try This! activities—one related to each section of the unit—that can be used to reinforce the unit's main concepts. The Try This! activities are located in the back of the Student Book.

Unit Focus

Packages and Polygons introduces students to the exploration of two- and three-dimensional shapes. The unit begins with an investigation into geometrical shapes, such as prisms and pyramids, in real-world objects. Students create nets and construct models of regular and semi-regular polyhedra. They use their models and nets to investigate the relationships between vertices, edges, and faces. Students solve problems by constructing, visualizing, and reasoning about polyhedra. At the end of the unit, students design their own packages in the shapes of polyhedra.

Mathematical Content

- investigating two-dimensional shapes
- investigating three-dimensional shapes (cube, cylinder, sphere, cone, triangular prism, rectangular solid, and pyramid)
- constructing three-dimensional models of geometric shapes
- applying Euler's formula
- identifying the properties of regular and semi-regular polyhedra
- investigating the stability of two- and three-dimensional shapes
- identifying the properties of Platonic solids
- investigating the relationships between edges, vertices, and faces

Prior Knowledge

This unit assumes students have an understanding of the following:
- relating two-dimensional representations to certain three-dimensional solids
- 90° and 45° angles
- the fact that the measure of a circle is 360°, and the measure of a straight line is 180°
- using a protractor or compass card to measure angles

Planning and Preparation

Pacing: 12 to 17 days

Section	Work Students Do	Pacing*	Materials
A. Packages	■ classify packages by different criteria, especially by shape ■ name common geometric shapes	1–2 days	■ Letter to the Family (one per student) ■ See page 5 of the Teacher Guide for a complete list of the materials and the quantities needed.
B. Nets	■ make nets of three-dimensional shapes ■ identify faces of a polyhedron ■ explore the height of a three-dimensional shape	1–2 days	■ See page 15 of the Teacher Guide for a complete list of the materials and the quantities needed.
C. Bar Models	■ construct bar models ■ identify and count vertices and edges ■ investigate the stability of bar models ■ reason about the structure of pyramids and prisms	2 days	■ See page 29 of the Teacher Guide for a complete list of the materials and the quantities needed.
D. Polygons	■ identify common polygons ■ establish criteria for identifying regular polygons ■ find the measures of interior and exterior angles of regular polygons	3 days	■ Student Activity Sheets 1–2 (one of each per student) ■ See page 39 of the Teacher Guide for a complete list of the materials and the quantities needed.
E. Platonic Solids	■ identify and name regular polyhedra ■ establish criteria for regularity ■ investigate the relationship between vertices, faces, and edges	2–3 days	■ Student Activity Sheets 3–9 (one of each per student) ■ See page 57 of the Teacher Guide for a complete list of the materials and the quantities needed.
F. Euler's Formula	■ apply Euler's formula to regular and other polyhedra	1–2 days	■ See page 71 of the Teacher Guide for a complete list of the materials and the quantities needed.
G. Semi-regular Polyhedra	■ study the relationship between regular and some semi-regular polyhedra, with emphasis on the structure of a soccer ball ■ construct models of five semi-regular polyhedra	2–3 days	■ Student Activity Sheets 10–13 (one of each per student) ■ See page 81 of the Teacher Guide for a complete list of the materials and the quantities needed.

* One day is approximately equivalent to one 45-minute class session.

Preparation

In the *Teacher Resource and Implementation Guide* is an extensive description of the philosophy underlying both the content and the pedagogy of the *Mathematics in Context* curriculum. Suggestions for preparation are also given in the Hints and Comments columns of this Teacher Guide. You may want to consider the following:

• Work through the unit before teaching it. If possible, take on the role of the student and discuss your strategies with other teachers.

• Use the overhead projector for student demonstrations, particularly with overhead transparencies of the student activity sheets and any manipulatives used in the unit.

• Invite students to use drawings and examples to illustrate and clarify their answers.

• Allow students to work at different levels of sophistication. Some students may need concrete materials, while others can work at a more abstract level.

• Provide opportunities and support for students to share their strategies, which often differ. This allows students to take part in class discussions and introduces them to alternative ways to think about the mathematics in the unit.

• In some cases, it may be necessary to read the problems to students or to pair students to facilitate their understanding of the printed materials.

• A list of the materials needed for this unit is in the chart on page xiii.

• Try to follow the recommended pacing chart on page xiii. You can easily spend much more time on this unit than the number of class periods indicated. Bear in mind, however, that many of the topics introduced in this unit will be revisited and covered more thoroughly in other *Mathematics in Context* units.

Resources

For Teachers	For Students
Books • Burton, D.M., *The History of Mathematics: An Introduction* (Dubuque, IA: Wm. C. Brown Publishers, 1991) • *Mathematics Assessment: Myths, Models, Good Questions, and Practical Suggestions,* edited by Jean Kerr Stenmark (Reston, Virginia: The National Council of Teachers of Mathematics, Inc., 1991)	MathSense Video • *The Same Again* (available from Encyclopædia Britannica)

Assessment

Planning Assessment

In keeping with the NCTM Assessment Standards, valid assessment should be based on evidence drawn from several sources. (See the full discussion of assessment philosophies in the *Teacher Resource and Implementation Guide*.) An assessment plan for this unit may draw from the following sources:

- Observations—look, listen, and record observable behavior.

- Interactive Responses—in a teacher-facilitated situation, note how students respond, clarify, revise, and extend their thinking.

- Products—look for the quality of thought evident in student projects, test answers, worksheet solutions, or writings.

These categories are not meant to be mutually exclusive. In fact, observation is a key part in assessing interactive responses and also a key to understanding the end results of projects and writings.

Ongoing Assessment Opportunities

- **Problems within Sections**
  To evaluate ongoing progress, *Mathematics in Context* identifies informal assessment opportunities and the goals that these particular problems assess throughout the Teacher Guide. There are also indications as to what you might expect from your students.

- **Section Summary Questions**
  The summary questions at the end of each section are vehicles for informal assessment (see Teacher Guide pages 12, 26, 36, 52, 68, 78, and 94).

End-of-Unit Assessment Opportunities

In the back of this Teacher Guide, there are five assessment activities that, when combined, can be completed in two 45-minute class sessions. For a more detailed description of these assessment activities, see the Assessment Overview (Teacher Guide pages 96 and 97).

In addition, students could write their own problems involving polygons, polyhedra, and Euler's formula. Students should also write the solutions, stating their assumptions and showing their calculations.

You may also wish to design your own culminating project or let students create one that will tell you what they consider important in the unit. For more assessment ideas, refer to the chart on pages xvi and xvii.

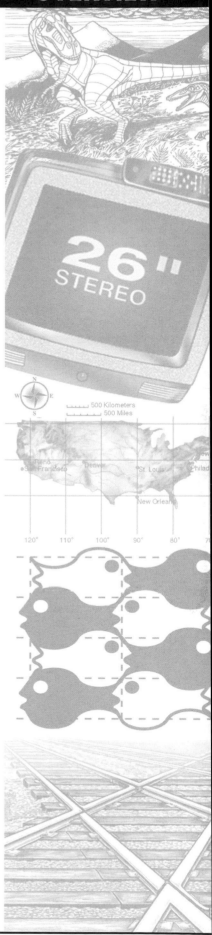

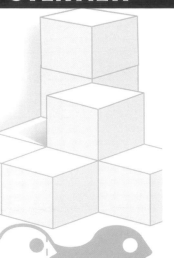

Goals and Assessment

In the *Mathematics in Context* curriculum, unit goals, categorized according to cognitive procedures, relate to the strand goals and to the NCTM Curriculum and Evaluation Standards. Additional information about these goals is found in the *Teacher Resource and Implementation Guide.* The *Mathematics in Context* curriculum is designed to help students develop their abilities so that they can perform with understanding in each of the categories listed below. It is important to note that the attainment of goals in one category is not a prerequisite to attaining those in another category. In fact, students should progress simultaneously toward several goals in different categories.

	Goal	Ongoing Assessment Opportunities	End-of-Unit Assessment Opportunities
Conceptual and Procedural Knowledge	**1.** recognize and identify geometrical shapes and structures in real objects and in representations	**Section A** p. 12, #6, #7 **Section G** p. 94, #11	Naming Shapes, p. 124 Polyhedra Constructions, p. 125 Alternative Assessment, p. 128
	2. understand and use Euler's formula	**Section F** p. 78, #6 p. 78, #7 **Section G** p. 94, #13	Tiling with Polygons, p. 126 Alternative Assessment, p. 128
	3. use the relationships between angles and turns to solve problems	**Section D** p. 50, #17	Tiling with Polygons, p. 126
	4. recognize and identify properties of regular polygons and polyhedra	**Section C** p. 36, #10 **Section D** p. 44, #10 **Section G** p. 94, #12	Naming Shapes, p. 124 Polyhedra Constructions, p. 125 Tiling with Polygons, p. 126 Alternative Assessment, p. 128
	5. construct geometric models	**Section C** p. 36, #10 **Section E** p. 60, #5 **Section G** p. 94, #11	Naming Shapes, p. 124 Alternative Assessment, p. 128
	6. draw two- and three-dimensional figures	**Section B** p. 22, #7 **Section C** p. 36, #10 **Section D** p. 44, #10 **Section F** p. 74, #2 p. 78, #6 **Section G** p. 94, #11	Naming Shapes, p. 124 Alternative Assessment, p. 128

	Goal	Ongoing Assessment Opportunities		End-of-Unit Assessment Opportunities
Reasoning, Communicating, Thinking, and Making Connections	**7.** make connections between different views of geometric solids	**Section B** **Section F** **Section G**	p. 22, #8 p. 74, #2 p. 94, #11	
	8. develop efficient counting strategies, involving geometric solids, that can be generalized	**Section E** **Section F**	p. 68, #11a p. 78, #7	Tiling with Polygons, p. 126 Hidden Faces, p. 127
	9. reason about the structure of the Platonic solids	**Section E**	p. 68, #10, #11b	Hidden Faces, p. 127 Alternative Assessment, p. 128

	Goal	Ongoing Assessment Opportunities		End-of-Unit Assessment Opportunities
Modeling, Nonroutine Problem-Solving, Critically Analyzing, and Generalizing	**10.** develop spatial visualization skills	**Section B** **Section C** **Section E** **Section F** **Section G**	p. 22, #8 p. 26, #13 p. 36, #10 p. 66, #9 p. 78, #6 p. 94, #11	Naming Shapes, p. 124 Polyhedra Constructions, p. 125 Hidden Faces, p. 127 Alternative Assessment, p. 128
	11. solve problems involving geometric solids	**Section B** **Section F**	p. 26, #13 p. 78, #7	Hidden Faces, p. 127 Alternative Assessment, p. 128

More about Assessment

Scoring and Analyzing Assessment Responses

Students may respond to assessment questions with various levels of mathematical sophistication and elaboration. Each student's response should be considered for the mathematics that it shows, and not judged on whether or not it includes an expected response. Responses to some of the assessment questions may be viewed as either correct or incorrect, but many answers will need flexible judgment by the teacher. Descriptive judgments related to specific goals and partial credit often provide more helpful feedback than percent scores.

Openly communicate your expectations to all students, and report achievement and progress for each student relative to those expectations. When scoring students' responses, try to think about how they are progressing toward the goals of the unit and the strand.

Student Portfolios

Generally, a portfolio is a collection of student-selected pieces that is representative of a student's work. A portfolio may include evaluative comments by you or by the student. See the *Teacher Resource and Implementation Guide* for more ideas on portfolio focus and use.

A comprehensive discussion about the contents, management, and evaluation of portfolios can be found in *Mathematics Assessment: Myths, Models, Good Questions, and Practical Suggestions,* pp. 35–48.

Student Self-Evaluation

Self-evaluation encourages students to reflect on their progress in learning mathematical concepts, their developing abilities to use mathematics, and their dispositions toward mathematics. The following examples illustrate ways to incorporate student self-evaluations as one component of your assessment plan.

- Ask students to comment, in writing, on each piece they have chosen for their portfolios and on the progress they see in the pieces overall.
- Give a writing assignment entitled "What I Know Now about [a math concept] and What I Think about It." This will give you information about each student's disposition toward mathematics as well as his or her knowledge.
- Interview individuals or small groups to elicit what they have learned, what they think is important, and why.

Suggestions for self-inventories can be found in *Mathematics Assessment: Myths, Models, Good Questions, and Practical Suggestions,* pp. 55–58.

Summary Discussion

Discuss specific lessons and activities in the unit—what the student learned from them and what the activities have in common. This can be done in whole-class discussion, in small groups, or in personal interviews.

Connections across the *Mathematics in Context* Curriculum

Packages and Polygons is the fifth unit in the geometry strand. The map below shows the complete *Mathematics in Context* curriculum for grade 7/8. It indicates where the unit fits in the geometry strand and in the overall picture.

A detailed description of the units, the strands, and the connections in the *Mathematics in Context* curriculum can be found in the *Teacher Resource and Implementation Guide.*

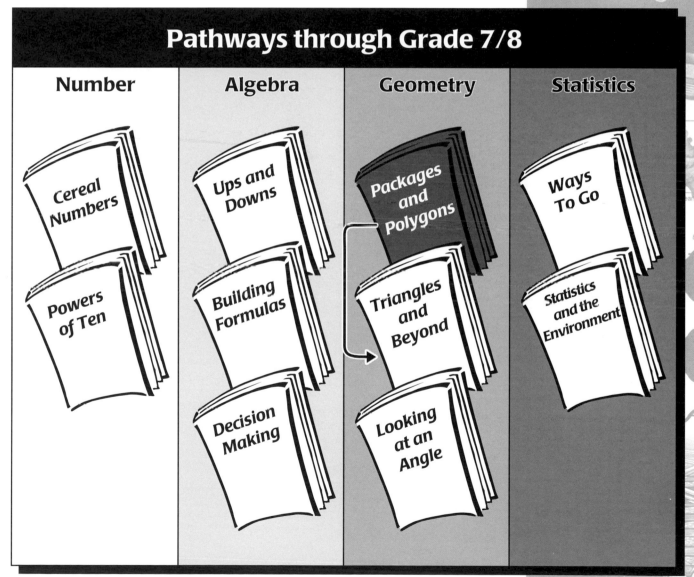

Pathways through Grade 7/8

Number	Algebra	Geometry	Statistics
Cereal Numbers	Ups and Downs	Packages and Polygons	Ways To Go
Powers of Ten	Building Formulas	Triangles and Beyond	Statistics and the Environment
	Decision Making	Looking at an Angle	

Pathways through the Geometry Strand
(Arrows indicate prerequisite units.)

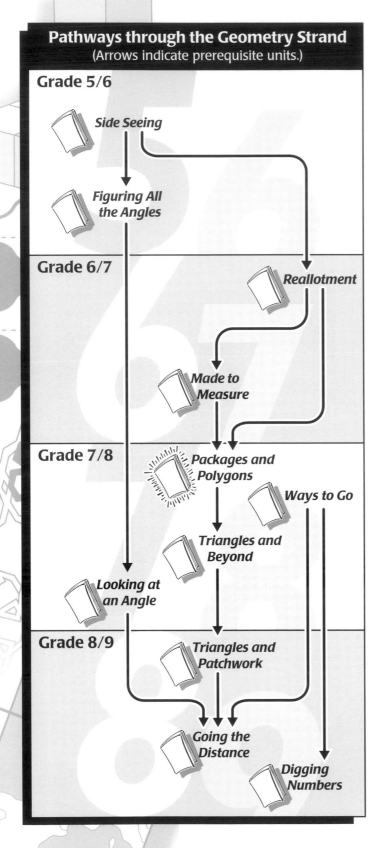

Grade 5/6

Side Seeing

Figuring All the Angles

Grade 6/7

Reallotment

Made to Measure

Grade 7/8

Packages and Polygons

Ways to Go

Triangles and Beyond

Looking at an Angle

Grade 8/9

Triangles and Patchwork

Going the Distance

Digging Numbers

Connections within the Geometry Strand

On the left is a map of the geometry strand; this unit, *Packages and Polygons,* is highlighted.

Packages and Polygons is the fifth unit in the geometry strand, and it is the first geometry unit in grade 7/8. Geometry in *Mathematics in Context* is presented through the concept of grasping space, to help students represent and make sense of the world. Through the exploration of real-world situations, students relate aspects of the real world to mathematical ideas. Understanding shapes is one of three substrands of the geometry strand. The exploration of space in the units in this strand and the more traditional geometry of forms and shapes blend together in *Packages and Polygons, Triangles and Beyond, Triangles and Patchwork,* and *Going the Distance.* In *Packages and Polygons,* students explore two- and three-dimensional shapes. They start by investigating geometrical shapes such as prisms and pyramids in real-world objects, such as packages. Students learn about characteristics of regular and irregular polygons and polyhedra by constructing paper and bar models, drawing two- and three-dimensional figures, and assembling and designing nets. They study the relationship among the numbers of edges, faces, and vertices of regular and semi-regular polyhedra, which leads to the use of Euler's formula.

The Geometry Strand

Grade 5/6

Side Seeing
Exploring the relationship between three-dimensional shapes and drawings of them, seeing from different points of view, and building structures from drawings.

Figuring All the Angles
Estimating and measuring angles and investigating direction, vectors, and rectangular and polar coordinates.

Grade 7/8

Packages and Polygons
Recognizing geometric shapes in real objects and representations, constructing models, and investigating properties of regular and semi-regular polyhedra.

Looking at an Angle
Recognizing vision lines in two and three dimensions; identifying and drawing shadows and blind spots; identifying the isomorphism of vision lines, light rays, flight paths, and so forth; understanding the relationship between angles and the tangent ratio; and computing with the tangent ratio.

Ways to Go
Reading and interpreting different kinds of maps, comparing different types of distances, progressing from one two-dimensional model to another (from a diagram to a map to a photograph to a graph), and drawing graphs and networks. (*Ways to Go* is also in the statistics strand.)

Triangles and Beyond
Exploring the interrelationships of the sides and angles of triangles as well as the properties of parallel lines and quadrilaterals, constructing triangles, and using transformations to become familiar with the concepts of congruence and similarity.

Grade 6/7

Reallotment
Measuring regular and irregular areas; discovering links between area, perimeter, surface area, and volume; and using English and metric units.

Made to Measure
Measuring length (including circumference), volume, and surface area using metric units.

Grade 8/9

Triangles and Patchwork
Understanding similarity and using it to find unknown measurements for similar triangles, and developing the concept of ratio through tessellation.

Going the Distance
Using the Pythagorean theorem to investigate distances, scales, and vectors and using slope, tangent, area, square root, and contour lines.

Digging Numbers
Using the properties of height, diameter, and radius to determine whether or not various irregular shapes are similar; predicting length using graphs and formulas; exploring the relationship between three dimensional shapes and drawings of them; and using length-to-width ratios to classify various objects. (*Digging Numbers* is also in the statistics strand.)

Connections with Other *Mathematics in Context* Units

Packages and Polygons is a grade 7/8 unit about understanding shapes in the geometry strand. The theme of shape and construction is the major focus of this unit. Shapes of common packages are used as a bridge to solid geometric figures. Students learned to visualize shapes from different views in *Side Seeing* and *Figuring All the Angles.* The theme of understanding where you are in space, what things look like, and how we create pictures of the three-dimensional world will be revisited in *Looking at an Angle.* Students have been introduced to angles and turns in the grade 5/6 unit *Figuring All the Angles.* The study of traditional angle relationships is formalized in *Triangles and Beyond.* Euler's formula gives students the opportunity to use a geometric formula that is supported by the approach to formulas students experienced in *Building Formulas.* The formula itself is revisited in *Patterns and Figures* in the algebra strand as a way to organize a series of number patterns.

The following mathematical topics that are included in the unit *Packages and Polygons* are introduced or further developed in other *Mathematics in Context* units.

Prerequisite Topics

Topic	Unit	Grade
angles, turns	*Figuring All the Angles***	5/6
patterns in numbers	*Patterns and Symbols****	5/6
different views of three-dimensional shapes	*Side Seeing***	5/6
	*Figuring All the Angles***	5/6
making two-dimensional representations of three-dimensional shapes	*Side Seeing***	5/6
formulas	*Expressions and Formulas****	6/7
	*Made to Measure***	6/7

Topics Revisited in Other Units

Topic	Unit	Grade
angles, turns	*Triangles and Beyond***	7/8
patterns in numbers	*Patterns and Figures****	8/9
different views of three-dimensional shapes	*Looking at an Angle***	7/8
making two-dimensional representations of three-dimensional shapes	*Looking at an Angle***	7/8
formulas	*Building Formulas****	7/8
	*Patterns and Figures****	8/9
vertices, edges, and faces	*Patterns and Figures****	8/9
polygons	*Triangles and Beyond***	7/8
	*Triangles and Patchwork***	8/9

** These units in the geometry strand also help students make connections to ideas about geometry.
*** These units in the algebra strand also help students make connections to ideas from geometry.

Student
Materials
and Teaching
Notes

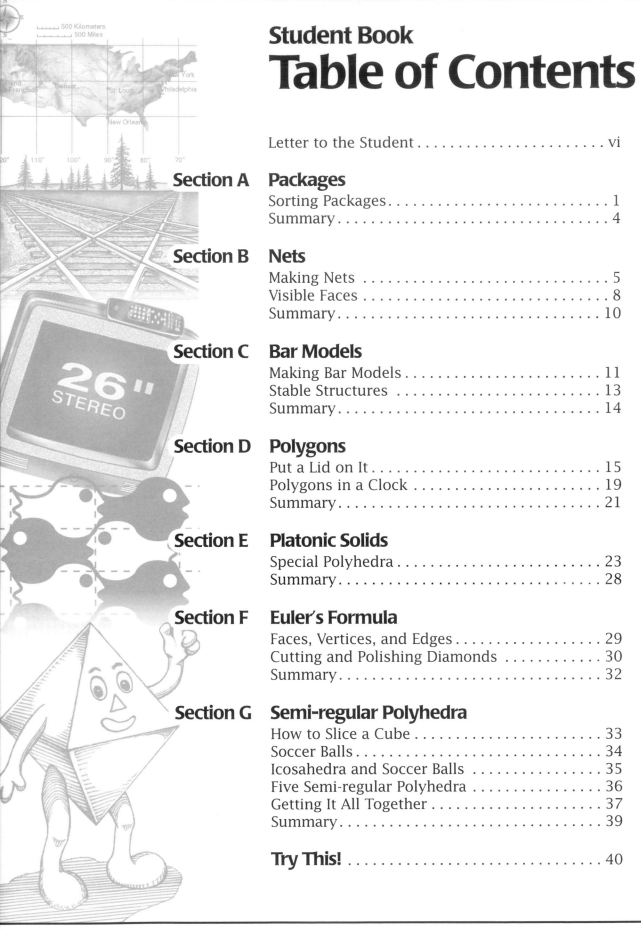

Student Book
Table of Contents

Dear Student,

Welcome to the unit *Packages and Polygons.*

Have you ever wondered why certain items come in differently shaped packages? The next time you are in a grocery store, look at how things are packaged. Why do you think table salt comes in a cylindrical package? Which packages do you think are the most practical?

Geometric shapes are everywhere. Look at the skyline of a big city. Can you see different shapes? Why do you think some buildings are built using one shape and some using another?

In this unit, you will explore a variety of two- and three-dimensional shapes and learn how they are related. You will build models of these

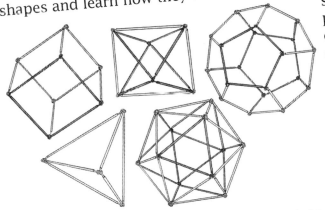

shapes using heavy paper, or straws and pipe cleaners, or gumdrops and toothpicks. As you work through the unit, notice the shapes of objects around you. Think about how the ideas you are learning in class apply to those shapes.

We hope you enjoy your investigations into packages and polygons.

Sincerely,

The Mathematics in Context Development Team

Work Students Do

Students find ways to classify packages and learn that packages can be classified according to shape. Students compare the shapes of three-dimensional models. They identify common geometric shapes in packages, in photographs, and in magazine illustrations. Students make a collage with labels for each shape. Finally, they classify pictures of common objects according to their shapes.

Goals

Students will:

- recognize and identify geometric shapes and structures in real objects and in representations.

Pacing

- approximately one or two 45-minute class sessions

Vocabulary

- truncate

About the Mathematics

This section introduces students to the following shapes: sphere, prism, right rectangular prism, cube, pyramid, cylinder, cone, and truncated cone. They informally describe properties of these shapes and find real-world examples of them. Students learn the difference between a prism, a pyramid, and a cone.

Materials

- Letter to the Family, page 110 of the Teacher Guide (one per student)
- packages of various shapes, page 7 of the Teacher Guide (at least five or six packages per group of students)
- newspapers and magazines, page 11 of the Teacher Guide (several per group of students)
- scissors, page 11 of the Teacher Guide (one pair per student)
- glue, page 11 of the Teacher Guide (one bottle per group of students)

Planning Instruction

You might begin the section by having a class discussion about packaging and why it is important. Ask students to start collecting packages of various shapes well in advance of the start of the section. Students could begin collecting a number of unusually shaped packages, such as cylindrical cans, rectangular and square boxes, heart-shaped containers of candy, milk cartons with a rectangular solid bottom and a triangular prism top, and other packages. You may need to supplement student collections, so it will help to have a classroom collection of packages available. These packages will be used throughout the unit.

Students may work on problems 1 and 3 in small groups. Problems 2, 4, and 5 can be done individually or in small groups. Students may work on problems 6 and 7 individually.

There are no optional problems in this section.

Homework

Problem 5 (page 10 of the Teacher Guide) can be assigned as homework. Also, the Bringing Math Home activity (page 11 of the Teacher Guide) can be assigned as homework. After students complete Section A, you may assign appropriate activities from the Try This! section, located on pages 40–43 of the *Packages and Polygons* Student Book. The Try This! activities reinforce the key mathematical concepts introduced in this section.

Planning Assessment

- Problems 6 and 7 may be used to informally assess students' ability to recognize and identify geometric shapes and structures in real objects and in representations.

A. PACKAGES

Sorting Packages

José did some shopping for a surprise party for his friends. When he got home, he put all the items he had bought on the table.

1. Discuss with your group the ways in which you might sort a collection like José's. Choose at least two different ways and explain how you would sort the collection.

Look around your home for some different-shaped packages. Select the shapes that you find the most interesting and bring them to class.

2. Pick one package from your collection or José's collection. Write a reason why the manufacturer chose that shape for the package.

Solutions and Samples
of student work

1. Answers will vary. Sample responses:

   - Sort by material
     glass: sugar dispenser, juice bottles, cola bottles
     steel/aluminum: container of cookies
     plastic: salt and pepper shakers
     cardboard: pizza box, donut box, hats

   - Sort by shape
     round sides: ball
     flat sides: cereal box, donut box, gift boxes
     round and flat sides: soft drink and juice bottles, cookie tins

   - Sort by type
     food: cookies and pizza
     toys: boxes, ball
     soft drinks: bottles

   Students might also suggest sorting into groups of food and toys or into groups of things in packages and things that are not packaged.

2. Answers will vary. Sample responses:

   Boxes and cans are easy to store.

   A perfume container might be designed so its appearance will attract attention and look nice.

Hints and Comments

Materials packages of various shapes (at least five or six packages per group of students)

Overview Students discuss different ways to sort a collection of packages and think of reasons why a manufacturer might choose a certain shape for a package.

Planning Have students bring a number of packages of various shapes to class. Students may work on problem **1** in small groups. You may wish to discuss this problem in class. Problem **2** can be done individually or in small groups.

Comments about the Problems

1. Sorting and classifying packages will help students to begin recognizing shapes and identifying properties of shapes. Allow students time to explore as many different ways of sorting the packages as possible. You may want to have students make charts to organize their data.

   Discuss with students why they might want to sort items in certain ways. For example, students might sort items by type of material in order to recycle them. Eventually, students should come up with the idea of sorting by shape. Some students may even know the names of different shapes. You might want to ask students to identify several shapes to find out how many names they already know.

2. You may want to encourage students to think about marketing reasons for various package shapes.

Look carefully at the shapes of your packages. Some shapes have special names.

3. Study the models shown below and on page 3.

    a. How are the three cylinders alike?

    b. Look at the truncated cones. What do you think truncated means?

    c. How are the prisms similar?

    d. What are some differences between a prism and a pyramid?

    e. What is the difference between a cone and a pyramid?

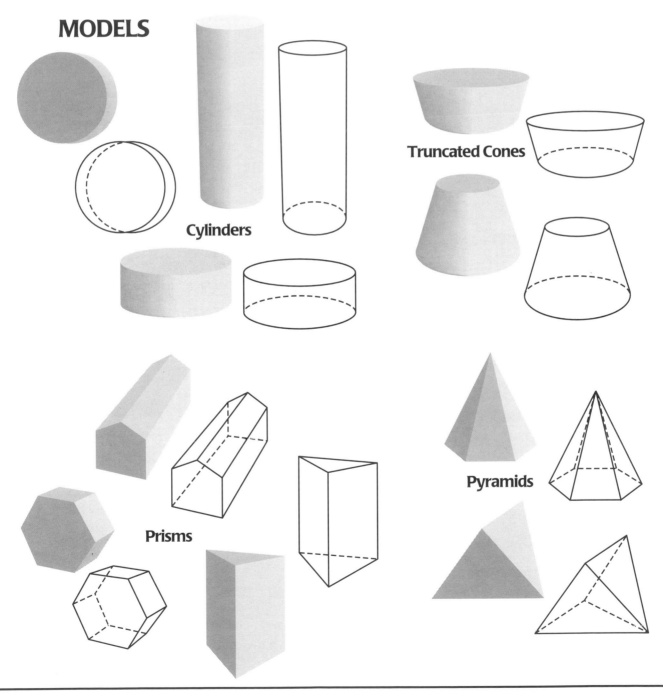

MODELS

Cylinders

Truncated Cones

Prisms

Pyramids

3. Answers will vary. Sample student responses:

a. The cylinders all have circles on the top and bottom.

b. Truncated means cut off.

c. Each of the prisms has at least two sides that are the same size and the same shape.

d. All the sides of a pyramid except the bottom come together in one point. The edges of the bottom side always stay at the same distance from the point at the top. A pyramid always has triangle-shaped sides, except the bottom. A prism has rectangle-shaped sides, except the top and bottom sides.

e. A pyramid looks like a cone with flat sides.

Overview Students learn the names of some three-dimensional shapes. In their own words, students describe differences and similarities between these shapes.

About the Mathematics At this stage, students may describe shapes using informal language. Later in the unit, mathematical language will be introduced. For example, what students may now refer to as a "side" is defined as a "face" on Student Book page 8.

Planning Students may work on problem **3** in small groups. However, you may want to discuss their answers as a class.

Comments about the Problems

3. Students learn to recognize three-dimensional shapes by describing them in different ways. Students may use informal language to describe the shapes, as long as they are able to make clear to each other what they mean. You may want to challenge them to be as precise as possible in their descriptions. Note that problem **3** also refers to the models at the top of page 3 of the Student Book (page 10 of the Teacher Guide).

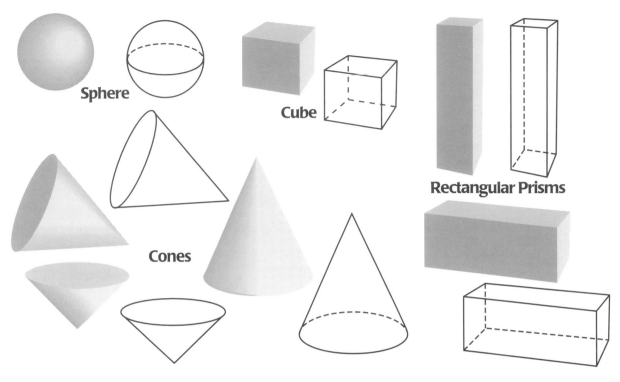

Sphere

Cube

Rectangular Prisms

Cones

4. List the packages in your collection and José's collection that are the following shapes:

    a. rectangular prism
        b. cube

    c. cylinder
        d. sphere

    e. cone
        f. truncated cone

    g. prism
        h. pyramid

Activity

Find pictures in newspapers and magazines of the shapes listed in problem **4.** Some pictures may include a combination of these shapes.

    5. Cut out the shapes and make a collage. Label each shape in your collage.

Note: At the end of this unit, you will design your own unique package. Keep this in mind as you work through the rest of the unit.

4. Answers will vary depending on the items in your classroom collection. Sample responses:

  a. box of chalk, cereal box, index card box

  b. box that a volleyball comes in

  c. soft drink can, soup can

  d. orange, grapefruit

  e. ice cream cone, party hat

  f. yogurt container, megaphone

  g. lightbulb box, cereal box, pizza box

  h. candy, puzzle box

5. Designs will vary.

Materials newspapers and magazines (several per group of students); scissors (one pair per student); glue (one bottle per group of students)

Overview Students find examples for each of eight listed shapes. Students make collages of shapes.

About the Mathematics Now that students have given an informal description of several three-dimensional shapes (on the previous page), they give examples of the shapes listed (on this page) to show that they know the properties of each shape.

Planning Students may work on problems **4** and **5** individually or in small groups. Problem **5** may be assigned as homework. You may want to create a display of students' collages in the classroom. Discuss students' solutions for problem **4** and their collages from problem **5.**

Comments about the Problems

  4. Make sure students give examples of each of the shapes listed. They may have to closely examine their collection of packages again and compare them to the models in the Student Book. Check students' answers. If students do not agree on the name of the shape of a certain package, you may want to review the names of shapes.

    d. You may want to discuss with students why they do not see many spherical packages.

  5. Homework This problem may be assigned as homework. Encourage students to find examples of as many different shapes as possible. It may be hard to find examples of some shapes. Ask students to make clear labels.

Bringing Math Home After the students' collages have been discussed in class and their labels checked, they can take their collages home and share them with their families. Students can explain to their families the relationship between each shape and its name. Together they can look for more examples of real-life packages and pictures of packages and label them appropriately.

Summary

In this section, you studied several three-dimensional shapes. You learned the difference between a prism, a pyramid, and a cone. You also learned that the word *truncate* means to shorten by cutting off.

Summary Questions

6. Look at the pictures below. What shape(s) do you recognize in each picture?

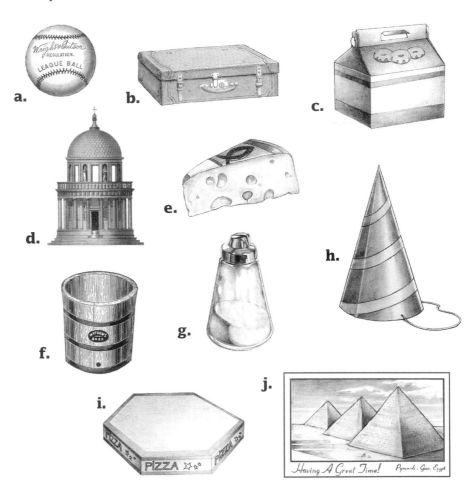

a.

b.

c.

d.

e.

h.

f.

g.

i.

j.

Having A Great Time! *Pyramids - Giza, Egypt*

7. a. Select three shapes that you studied in this section. Write down a characteristic shared by two of the shapes, but not by the third.

b. Repeat part **a** with three different shapes.

6. a. baseball   sphere

   b. suitcase   rectangular prism

   c. donut box   prism

   d. domed building   cylinder; half a sphere on top

   e. wedge of Swiss cheese   prism

   f. barrel   cylinder

   g. sugar container   truncated cone

   h. party hat   cone

   i. pizza box   prism

   j. Egyptian pyramids   pyramid

7. Answers will vary. Sample student responses:

   a. A pyramid, a prism, and a sphere—the pyramid and the prism have edges that are straight lines and the sphere does not have any straight edges.

   b. A cube, a prism, and a cylinder—the cube and prism have only flat sides, but the cylinder has a round side as well.

Overview Students read the Summary, which reviews the main concepts covered in this section. They label pictures with names of shapes and describe and compare characteristics of those shapes.

Planning Students may work on problems **6** and **7** individually. These problems may be used as an informal assessment. After students complete Section A, you may assign appropriate activities from the Try This! section, located on pages 40–43 of the *Packages and Polygons* Student Book, for homework.

Comments about the Problems

  6–7. Informal Assessment These problems assess students' ability to recognize and identify geometric shapes and structures in real objects and in representations.

      Discuss students' answers in class. Focus the discussion on the differences between the shapes. You may want to conclude the discussion by creating a list of informal definitions of the shapes studied in this section.

SECTION B. NETS

Work Students Do

In this section, students cut apart milk cartons or boxes to make two-dimensional nets. They visualize what three-dimensional shapes could be made by folding two-dimensional nets. Students are introduced to the concept of height in a three-dimensional shape. They identify how many faces of a polyhedron are visible and how many are hidden in given views. From an assortment of nets, students choose the one that could be folded to make a box. Finally, students visualize how each side of a net must be painted, in order for it to be folded to make a cube that has opposite sides that are the same color.

Goals

Students will:

- construct geometric models;*
- draw two- and three-dimensional figures;
- make connections between different views of geometric solids;
- develop spatial visualization skills;
- solve problems involving geometric solids.

  * This goal is introduced in this section and is assessed in other sections of the unit.

Pacing

- approximately one or two 45-minute class sessions

Vocabulary

- face
- net

About the Mathematics

Nets are a way of visualizing solids and help in identifying properties of solids. A net is a two-dimensional figure that can be folded into a three-dimensional solid. The visualizing aspect of this section is important. A sense of shape depends to a large extent on mental images. For instance, any two-dimensional picture of a three-dimensional solid will require the viewer to supply information about the parts of the solid that are hidden. Students will learn that the properties of the two-dimensional shape (such as the length and width of each face) are extended to three dimensions (such as height) when the net is folded. Using nets as models enables students to create definitions of, to conjecture about, and to gain insight into geometric relationships.

Materials

- small milk cartons or shoe boxes, page 17 of the Teacher Guide (two per pair of students)
- scissors, pages 17, 19, 21, 23, 25, and 27 of the Teacher Guide (one pair per pair or group of students)
- centimeter rulers, pages 17, 19, 21, 23, and 25 of the Teacher Guide (one per student)
- transparency of a box, page 17 of the Teacher Guide, optional (one per class)
- overhead projector, page 17 of the Teacher Guide, optional (one per class)
- small box, page 17 of the Teacher Guide, optional (one per class)
- cube-shaped objects, page 23 of the Teacher Guide (one per student)
- tracing paper, pages 19, 21, and 27 of the Teacher Guide, optional (one sheet per student)
- graph paper, page 25 of the Teacher Guide, optional (one sheet per student)
- colored pencils, page 27 of the Teacher Guide, optional (three per student)
- glue or tape, pages 19, 21, 23, 25, and 27 of the Teacher Guide, optional (one dispenser per group of students)

Planning Instruction

You may want to begin this section by discussing how packages are made. Ask students if they have ever folded or unfolded a preformed cardboard box. Ask students, *Why is it helpful to be able to fold up a box when it is not being used?* [Boxes take up less storage space when they are folded flat.]

Students may work on problems 1 and 2 in pairs. The remaining problems may be done individually or in small groups.

There are no optional problems in this section.

Homework

Problems 10, 11, and 12 (page 24 of the Teacher Guide) may be assigned as homework. Also, the Extension (page 27 of the Teacher Guide) can be assigned as homework. After students complete Section B, you may assign appropriate activities from the Try This! section, located on pages 40–43 of the *Packages and Polygons* Student Book. The Try This! activities reinforce the key mathematical concepts introduced in this section.

Planning Assessment

- Problem 7 may be used to informally assess students' ability to draw two- and three-dimensional figures.
- Problem 8 may be used to informally assess students' ability to make connections between different views of geometric solids and to develop spatial visualization skills.
- Problem 13 may be used to informally assess students' ability to develop spatial visualization skills and to solve problems involving geometric solids.

B. NETS

Making Nets

Activity

Cut the top off a milk carton or a small box. Cut along the edges of the carton to open it up and lay it flat.

Cut another carton or box a different way. Open it up and lay it flat.

The flat patterns you made from the cartons or boxes are called nets.

1. **a.** Compare the two nets you made. Do they look the same? If not, what is the difference?

   b. Draw the two nets that you made.

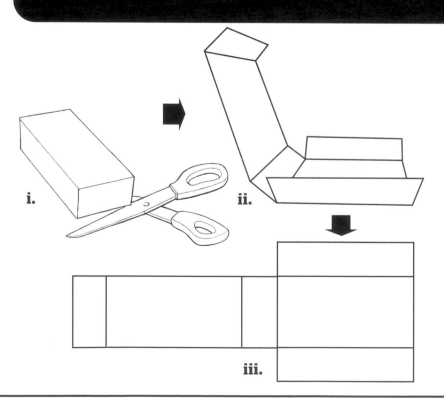

i.

ii.

iii.

2. **a.** How was the box pictured in step i on the left cut to make the net that is shown in step iii?

   b. If the net shown is the actual size, what is the length, width, and height of this box in centimeters?

1. a. No, the nets are not the same. For examples, see the solutions to problem **1b.** To create net I, for example, the box was cut along the upright four edges, down to the base. All four sides are cut loose from each other. To create net II, there are only two cuts, along two upright edges, down to the base, and then continuing along the base. Three of the upright sides of the box are still attached to each other.

b. The nets should look different. Sample drawings:

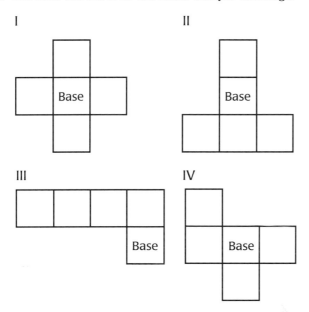

I   II

III   IV

2. a. It was cut along the seven marked edges to flatten it out as shown below.

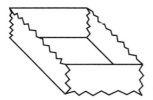

b. The length = 3.6 cm, the width = 1.9 cm, and the height = 0.9 cm.

Materials small milk cartons or shoe boxes (two per pair of students); scissors (one pair per pair of students); centimeter rulers (one per student); transparency of a box, optional (one per class); overhead projector, optional (one per class); small box, optional (one per class)

Overview Students cut apart milk cartons or boxes to make different two-dimensional nets. They draw the nets they made and compare them.

Planning Students may work on problem **1** in pairs. However, you may want to discuss their answers as a class.

Comments about the Problems

1. Make sure students rinse the milk cartons before bringing them to class. Before students cut open the cartons, you may wish to have them cut off the top of the carton in such a way that a cube-shaped box is the result.

Be sure students understand that in order to create a net, all sides of the box have to be attached in some way. There should be no loose parts. Remind students that the top face will be missing. Students should try to cut apart the second carton in a different way.

Compare the different nets that can be made from this one shape. Some students will be able to visualize folding a net. For others this will be more difficult. Let students guess what will be the result of folding (or unfolding) and let them check their guesses by actually doing the folding or unfolding.

Ask students, *What makes nets different?* [They are different in the way they are created. For each different net, the cuts are made along different edges. So different sides remain attached in the resulting net.]

2. a. You may want to make a transparency of the box so students can demonstrate where it was cut. For some students who have difficulty visualizing, it may be necessary to actually cut open a box similar to the one shown.

3. What shape would each net pictured below make if it were folded up?

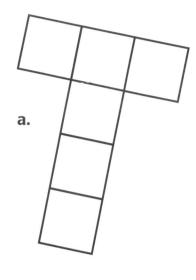

a.

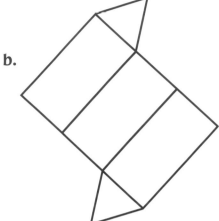

b.

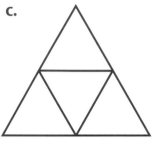

c.

4. Explain what would happen if you folded the net shown below.

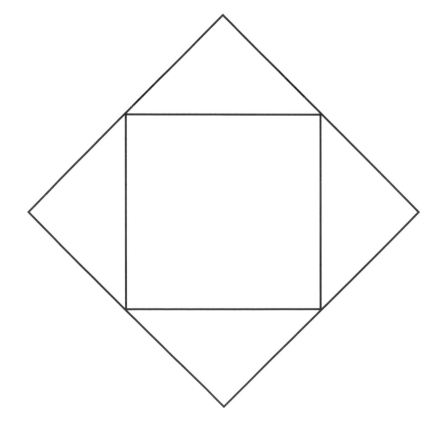

3. **a.** a cube

   b. a (triangular) prism

   c. a pyramid with four faces

4. The triangular faces do not touch until they have fallen flat. The faces are too short to construct a pyramid.

   Some students may say that each triangular face folds over the square and completely covers it. Other students may say it folds flat as shown below.

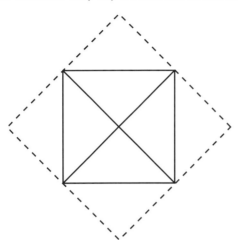

Materials tracing paper, optional (one sheet per student); scissors, optional (one pair per group of students); centimeter rulers, (one per student); glue or tape, optional (one dispenser per group of students)

Overview Students visualize folding up four given nets. Then they describe the shape each net would make.

Planning Students may work on problems **3** and **4** individually or in small groups.

Comments about the Problems

3. Encourage students to visualize the folding process. Some students may have to make the nets (using tracing paper to trace the shapes from the Student Book and scissors to cut them out) and fold them in order to see the shape.

4. Again, encourage students to answer the question without actually making and folding the net. Students may use a centimeter ruler and do whatever measuring they think is necessary.

5. Which of the following nets can be folded to make a pyramid? Explain.

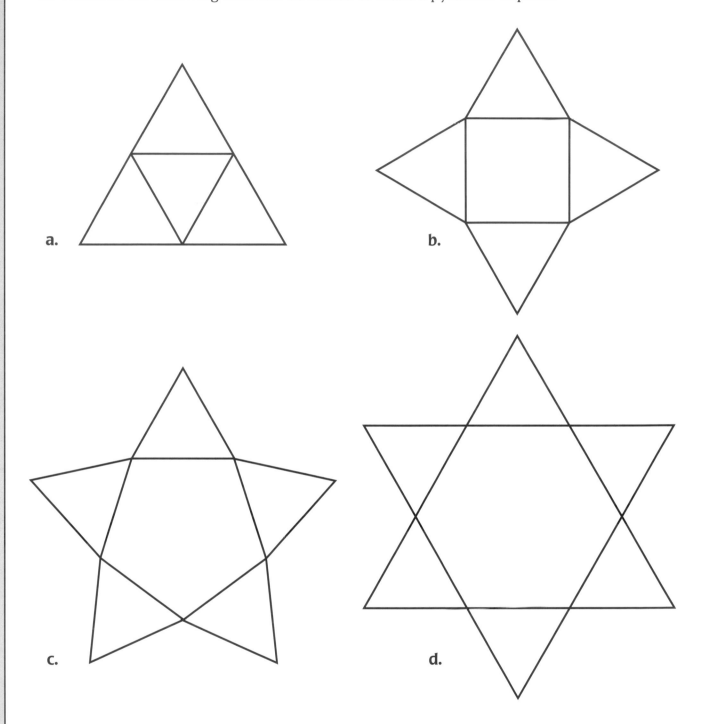

a.

b.

c.

d.

6. a. Sandra says, "The pyramids you can make from the above nets all have the same height." Is this true? Why or why not?

b. Describe how you can measure the height of a pyramid.

5. Nets a, b, and c can be folded into pyramids. The sides of the triangles in net d are not long enough. The triangular faces are too small to make a three-dimensional shape. When folded, this net makes a two-dimensional shape.

6. a. This statement is not true. Nets a, b, and c will make pyramids of different heights. Net d will not make a pyramid.

   b. Answers will vary. Sample response:
   You may put the pyramid on a table and then put a ruler straight up to see how high the pyramid is:

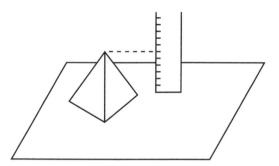

Materials tracing paper, optional (one sheet per student); scissors, optional (one pair per group of students); centimeter rulers (one per student); glue or tape, optional (one dispenser per group of students)

Overview Students visualize folding four given nets to predict which can be folded into a pyramid. They also think about ways to determine the height of a pyramid.

About the Mathematics The distance from the vertex on any face of a regular pyramid to the opposite edge is called the slant height. The height or altitude of a pyramid is the distance from the vertex to the base.

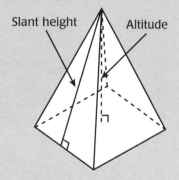

Slant height   Altitude

Planning Students may work on problems **5** and **6** individually or in small groups. Discuss their answers, especially to problem **6.**

Comments about the Problems

   5. Again, as in problem **3,** encourage students to visualize folding the nets. Still, some students may have to trace the net on paper, cut it out, and actually fold it to see what happens.

   6. a. Some students might suggest that Sandra is correct because the faces of the outer triangles are the same. They may have to build the nets to see that the length of the sides does not determine height.

      b. Students may have different methods. Some students may suggest measuring the distance from the top of the pyramid to the base using a ruler or string.

Visible Faces

Find an object that is shaped like a cube. Each flat side of a cube is called a *face*.

7. a. Hold the cube so that you see only one face. Draw what you see.

b. Hold the cube so that you see exactly two faces. Draw what you see.

c. Hold the cube so that you see exactly three faces. Draw what you see.

d. What happens when you try to hold the cube so that you can see four faces?

In a net, all faces of a shape are visible.

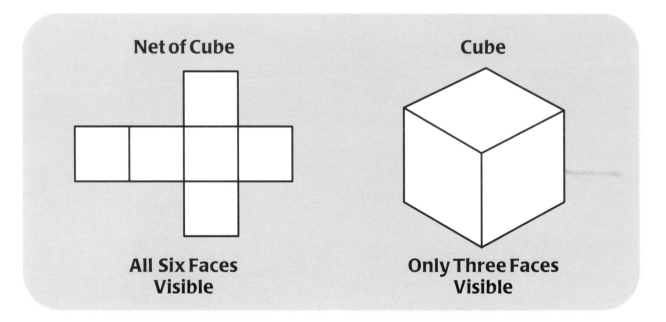

Net of Cube

All Six Faces Visible

Cube

Only Three Faces Visible

8. Look at the prism on the right, only two faces are visible.

a. How many faces are hidden?

b. How many faces are triangles? How many faces are rectangles?

c. Draw a net of this prism.

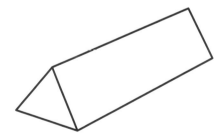

9. The object on the right is a prism, too. It looks like a bolt.

a. How many faces are hidden?

b. Draw a net of this shape.

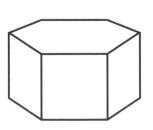

7. a.

b.

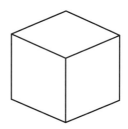

c. Drawings will vary depending on eye level. One possibility:

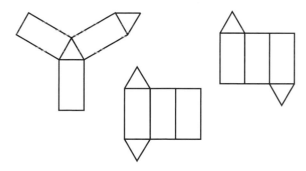

d. It is not possible to hold the cube to see four faces.

8. a. Three faces are hidden.

b. The prism has two triangular faces and three rectangular ones.

c. Drawings will vary. Three possibilities:

9. a. Four faces are hidden.

b. Drawings will vary. One possibility:

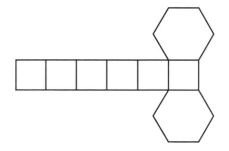

Materials centimeter rulers (one per student); cube-shaped objects with six flat sides (one per student); scissors (one pair per group of students); glue or tape, optional (one dispenser per group of students)

Overview Students draw views of a cube showing different numbers of faces. They visualize hidden faces of prisms and draw nets of prisms.

About the Mathematics Describing what an object looks like from different views without actually seeing it requires visualization skills. So does drawing the nets of objects of which you see only one view.

Planning Students may work on problems **7–9** individually or in small groups. Problems **7** and **8** may be used as informal assessments. Discuss problem **7** as a class.

Comments about the Problems

7. **Informal Assessment** This problem assesses students' ability to draw two- and three-dimensional figures.

   You may have to help some students draw the faces. Point out that the vertical lines are all parallel. Students might try looking at the cube with one eye closed to make it easier to draw what they see.

8. **Informal Assessment** This problem assesses students' ability to make connections between different views of geometric solids and develop spatial visualization skills.

   c. Challenge the students to find more than one possibility.

9. Like problem **8,** this problem requires visualization skills. Encourage students to answer the problem by looking at the pictures only. For students having difficulty visualizing, you may want to have an actual bolt available.

   b. Students may check the correctness of their nets by cutting them out and folding them.

Extension You may have students hold other shapes and determine how many sides they can see.

The drawing shown below is of
a cube-shaped box without a top.

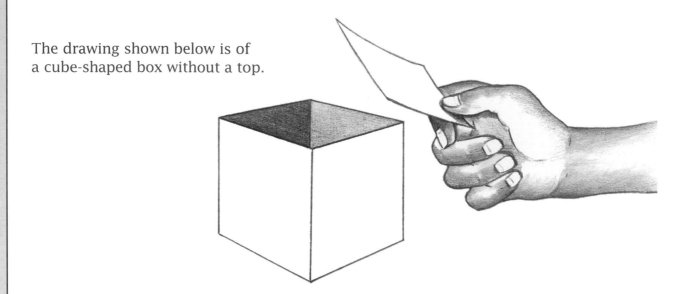

10. Which of these nets can be folded to make the above box?

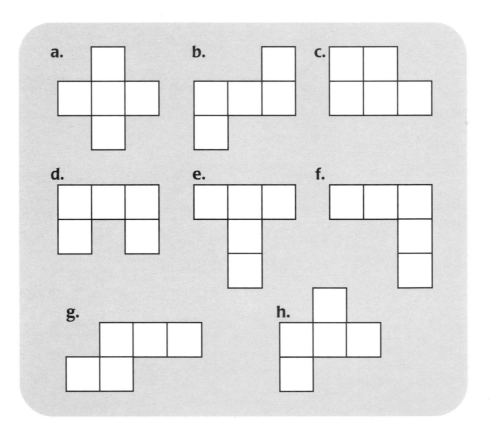

a.

b.

c.

d.

e.

f.

g.

h.

11. On a sheet of graph paper, trace each net that you selected for problem **10.**
On each net, shade the face that is the bottom, or base, of the box.

12. There are three more nets that could make the above box. Draw at least one
of these nets and shade the bottom face.

10. a. a, b, e, g, and h

11. a.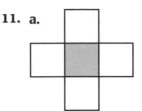

b.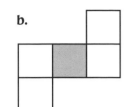

e. 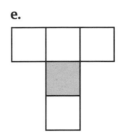 **g.**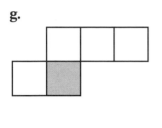

h.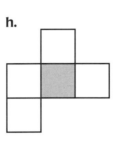

12. Drawings will vary. Three possibilities:

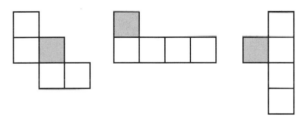

Materials graph paper, optional (one sheet per student); centimeter rulers (one per student); scissors, optional (one pair per group); glue or tape, optional (one dispenser per group)

Overview Students determine which nets can be folded into an open cube and indicate which is the bottom face.

Planning Students may work on problems **10–12** individually or in small groups. These problems may also be assigned as homework. Discuss students' answers.

Comments about the Problems

10–12. Homework These problems may be assigned as homework.

10. Encourage students to answer the problems only by looking at the nets.

11. Some students may have to make small models in order to determine which face is the bottom. Models can be made using graph paper. Students may have created nets of cubes in the activity for problem **1,** depending on the shape of the container they cut apart.

12. Students may check their answers by cutting out the nets and folding them. Some students may think that they can find even more nets to make the box. Discuss whether or not these nets would be different. They may be rotations or reflections of the nets that were already found.

Summary

A *net* is a flat pattern that forms a three-dimensional shape when folded. You can see all the faces of a shape in a net. You can also make different nets for the same shape. Nets can be useful in solving problems about shapes.

Summary Questions

A cube is painted red on two faces, white on two faces, and with a blue dot on two faces.

Opposite faces are painted in the same way.

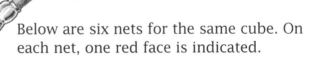

Below are six nets for the same cube. On each net, one red face is indicated.

13. Trace the nets in your notebook and mark the correct color for each of the other faces.

a.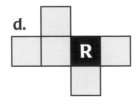

b.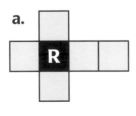

c.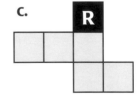

d.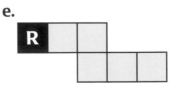

e.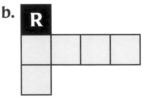

f.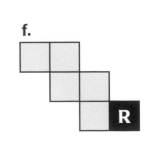

13. Answers will vary. Sample responses:

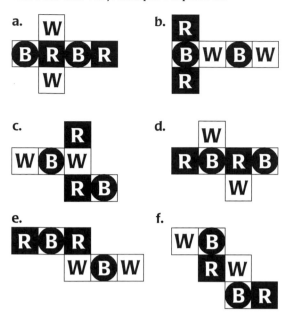

a.

b.

c.

d.

e.

f.

Other solutions can be found by exchanging blue and white.

Materials tracing paper, optional (one sheet per student); scissors, optional (one pair per group); colored pencils, optional (three per student); glue or tape, optional (one dispenser per group)

Overview Students read the Summary, which reviews the main concepts covered in this section. They also work on a problem involving nets of a cube in which the faces are identified as having different colors.

Planning Have students read and discuss the Summary. Problem **13** may be used as an informal assessment. Students may work on this problem individually or in small groups. Discuss their answers. After students complete Section B, you may assign appropriate activities from the Try This! section, located on pages 40–43 of the *Packages and Polygons* Student Book, for homework.

Comments about the Problems

13. **Informal Assessment** This problem assesses students' ability to develop spatial visualization skills and to solve problems involving geometric solids.

   Some students might need to make and fold nets to do this problem. It is important to note that the blues and whites are interchangeable.

   If students are having difficulty visualizing what shape is formed from a net, encourage them to think about the shape first, and then make and fold the net to check their answers.

Extension You may have students draw nets of number cubes. They worked with number cubes in the grade 5/6 unit *Take a Chance*. Note that the numbers on opposite faces of a number cube add up to 7 (1 is opposite 6, and so on).

SECTION C. BAR MODELS

Work Students Do

In this section, students construct bar models of polyhedra. They identify and count the vertices and edges on their bar models and test their models for stability. Students learn by investigating that triangles are very stable figures. Students use this information to build a structure that is strong enough to hold a book. Throughout the section, students answer questions and solve problems regarding the structure of pyramids and prisms.

Goals

Students will:

- recognize and identify properties of regular polygons and polyhedra;
- construct geometric models;
- draw two- and three-dimensional figures;
- develop spatial visualization skills.

Pacing

- approximately two 45-minute class sessions

Vocabulary

- edge
- vertex

About the Mathematics

Bar models of polyhedra can be studied to find the properties of polyhedra.

The focus of the activities in this section is to have students become familiar with different polyhedra and how the number of vertices, faces, and edges differ for each. Each vertex of a face is a vertex of the polyhedron. Each side of a face is an edge of the polyhedron. A useful property of the tetrahedron is that it is stable or rigid. This is the reason that it is used so often in the construction of bridges.

Materials

- gumdrops or modeling clay, pages 31, 33, 35, and 37 of the Teacher Guide (see specific pages for quantities needed)
- toothpicks, straws, or pipe cleaners, pages 31, 33, 35, and 37 of the Teacher Guide (see specific pages for quantities needed)
- centimeter rulers, pages 33 and 37 of the Teacher Guide (one per student)
- scissors, page 35 of the Teacher Guide (one pair per student)
- glue or tape, page 35 of the Teacher Guide (one dispenser per pair of students)
- posterboard or cardboard, page 35 of the Teacher Guide (several sheets per pair of students)
- paper cup, page 35 of the Teacher Guide (one per class)
- quarters, page 35 of the Teacher Guide (several per class)

Planning Instruction

You may want to begin this section with a short class discussion about how polyhedra are often used in building structures. Students could name examples of buildings and bridges in their community or structures of objects around the school that have the properties of a polyhedron.

Students may work on problem 1 in small groups. Students may work individually or in small groups on problems 2–5. Problems 6, 8, 9 and 10 may be done individually. Students may work on problem 7 in pairs.

Problem 7 is optional. If time is a concern, you may omit this problem or assign it as homework.

Homework

Problems 6 and 8 (page 34 of the Teacher Guide) and 9 (page 36 of the Teacher Guide) can be assigned as homework, if students have the necessary materials available. Also, the Extension (page 37 of the Teacher Guide) can be assigned as homework. After students complete Section C, you may assign appropriate activities from the Try This! section, located on pages 40–43 of the *Packages and Polygons* Student Book. The Try This! activities reinforce the key mathematical concepts introduced in this section.

Planning Assessment

- Problem 10 may be used to informally assess students' ability to recognize and identify properties of regular polygons and polyhedra. It can also be used to informally assess students' ability to construct geometric models, draw two- and three-dimensional figures, and develop spatial visualization skills.

C. BAR MODELS

Activity

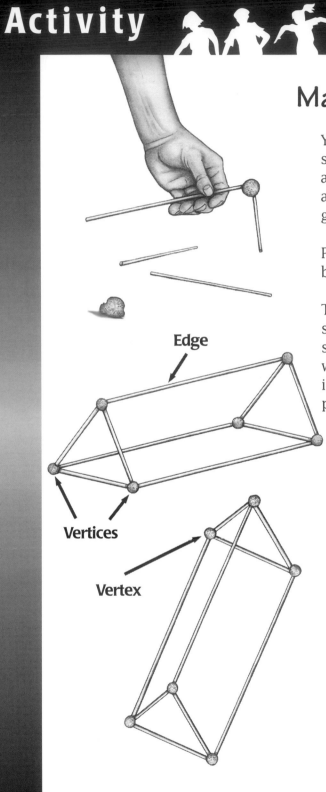

Making Bar Models

You can make a bar model of a shape by using drinking straws and pipe cleaners, toothpicks and clay, or toothpicks and gumdrops.

Pictured below on the left is a bar model of a triangular prism.

The bars are the *edges* of the shape (the lines that form the shape). A point at a corner where two or more edges meet is called a *vertex*. (*Note:* The plural of vertex is vertices.)

Edge

Vertices

Vertex

1. Make a bar model of a cube.

   a. How many vertices and how many edges does the model have?

   b. Is this bar model stable? Why or why not?

1. a. The model has 8 vertices and 12 edges.

b. Answers will vary. Sample response:

No, the bar model is not stable. You can make it change shape.

Instability of a Cube

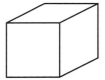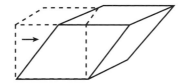

Materials gumdrops or modeling clay (10 pieces per student); toothpicks, straws, or pipe cleaners (20 per student)

Overview Students make a bar model of a cube and judge its stability.

About the Mathematics This is the first exploration of the fact that the shape of a triangle is determined by its sides and the shape of a quadrilateral is not. A triangle has one and only one shape for a given set of sides. A quadrilateral can change shape. *Stable* means fixed or firm, not likely to give way.

Planning This can be a time-consuming section because of the hands-on activities. To save time, it may be helpful to have groups of students work together to make the bar models instead of having each student individually make every model. Every student should have the opportunity to make at least one bar model. Keep the student models to use as a point of reference for the rest of the unit. Students may work on problem **1** in small groups. Discuss this problem with the class.

Comments about the Problems

1. a. Make sure students are familiar with the terms *vertices* and *edges* so that they will be able to count the number of each.

b. In this section, students will discover that some structures are more stable or stronger than others.

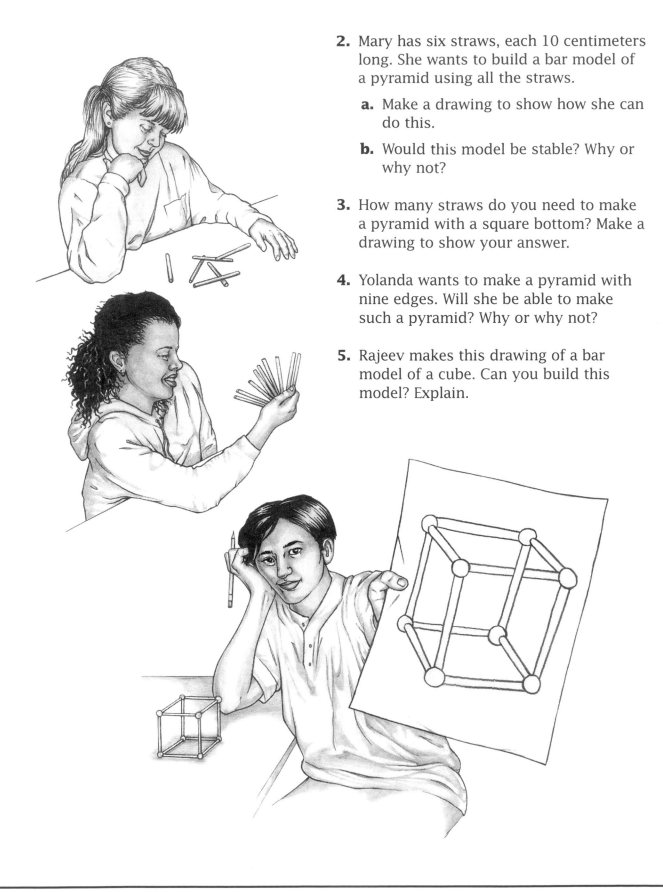

2. Mary has six straws, each 10 centimeters long. She wants to build a bar model of a pyramid using all the straws.

   a. Make a drawing to show how she can do this.

   b. Would this model be stable? Why or why not?

3. How many straws do you need to make a pyramid with a square bottom? Make a drawing to show your answer.

4. Yolanda wants to make a pyramid with nine edges. Will she be able to make such a pyramid? Why or why not?

5. Rajeev makes this drawing of a bar model of a cube. Can you build this model? Explain.

2. a.

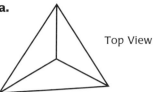

Top View

b. Yes. It will not collapse easily. It maintains its structure or shape.

3. Eight straws are needed (four for making the base, and four for connecting the base with the top).

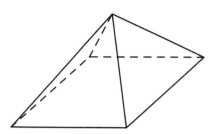

4. No. Yolanda can make a pyramid with eight edges (and have one left over) or make a pyramid with 10 edges (if she uses one extra edge).

5. No. Rajeev has drawn an impossible model. Edge 1 should be behind edge 2, as shown below.

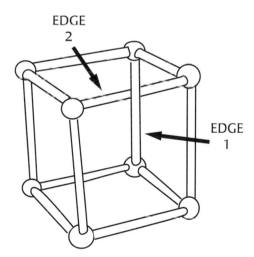

EDGE 2

EDGE 1

Materials centimeter rulers (one per student); gumdrops or modeling clay, optional (30 pieces per student); toothpicks, straws, or pipe cleaners, optional (40 per student)

Overview Students investigate the number of straws needed to make different pyramids. They are also confronted with an impossible figure.

Planning You may want to have students work on problems **2–5** individually or in small groups. Discuss students' solutions and strategies for each problem.

Comments about the Problems

2. a. Challenge students to answer the question without constructing the pyramid. If they have difficulty, they can build it. Remind students that there are many different angles from which they can draw the pyramid. Ask students, *What side of the pyramid are you viewing?* and *Where is the pyramid's vertex from this view?*

b. The pyramid is stable because it is made of triangles. This structure is often used in buildings because of its stability. The pyramid can be strengthened by adding two diagonals to its square face as shown below.

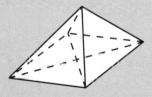

3. If students have difficulty with this problem, they may construct the pyramid before making a drawing of it.

4. The number of edges of a pyramid is always even. There are as many edges in the base as there are edges in the top. Some students may figure out by reasoning that a pyramid must have an even number of edges. Other students may need to try building a pyramid with an odd number of edges first. (Students added odd and even numbers in the unit *Patterns and Symbols*.)

5. *Belvedere,* a famous picture by M. C. Escher, depicts this strange cube. If you can find *Belvedere,* discuss this picture with your class. (*The World of M. C. Escher.* New York: Harry N. Abrams, Inc. Publishers, 1971.)

Stable Structures

Activity

6. Build a bar model with five vertices and eight edges. Make another with six vertices and nine edges.

    a. Which structure is more stable? How does the number of triangular faces affect the stability of the structure?

    b. What could you do to make the other structure more stable?

    c. Is it possible to build a structure with nine vertices and five edges? Explain.

7. Make a structure that is strong enough to hold a book using 14 vertices and 20 edges.

To make a framework rigid, construction workers build triangles within the framework.

8. Suppose a carpenter makes the bookcase shown on the right.

    a. Do you think this is a good design? Why or why not?

    b. You can improve this design by adding one bar. Where would you place this bar?

6.

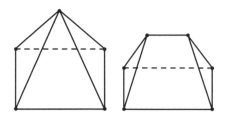

a. The structure with five vertices and eight edges is more stable because more of the faces are triangles.

b. add a diagonal on the quadrilateral base

c. It is impossible to build a closed structure with more vertices than edges. In a closed structure, every edge is connected to every other edge through a vertex. This means that each vertex must have more than one edge attached to it.

7. Structures will vary. Sample structure:

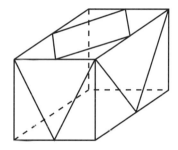

8. a. No, the bookcase is not stable. It can easily be moved left or right. This can be demonstrated by using a simple cardboard model as shown below.

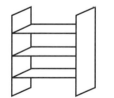 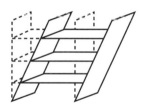

b. One diagonal bar is sufficient to make the construction rigid because the diagonal bar creates triangles within the framework as shown below.

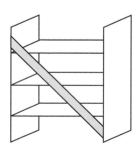

Materials gumdrops or modeling clay (100 pieces per student); toothpicks, straws, or pipe cleaners (200 per student); scissors (one pair per student); glue or tape (one dispenser per pair of students); posterboard or cardboard (several sheets per pair of students); paper cup (one per class); quarters (several per class)

Overview Students construct bar models and investigate their stability. They also judge the stability of a given design and try to improve upon it.

About the Mathematics Triangles and tetrahedrons increase the stability of structures.

Planning Problems **6** and **8** may be assigned as homework. You may want to discuss students' answers in class. Students may work on problem **7** in pairs. However, if time is a concern, you may omit this problem or assign it as homework.

Comments about the Problems

6. Homework This problem can be assigned as homework. Stress that the gumdrops are the vertices and the toothpicks are the edges. You may want to remind students that adding a vertex anywhere but at the end of an edge is not allowed. Discuss why some structures are more stable than others.

7. This problem is optional. Students may use the information they discovered in problem **6,** that triangles and tetrahedrons increase stability. If students have difficulty, they may make the structure with fewer vertices and edges.

8. Homework This problem can be assigned as homework. Encourage students to make a cardboard model. This will give them the opportunity to experiment with ways to make the bookcase stable.

Extension Students may work in small groups to build a bridge between two chairs that are 30 centimeters (about 1 foot) apart. Have them use as many vertices and edges as they think necessary. When they are done, see if the bridge will hold a paper cup with five quarters in it. You might want to have a class competition to see which group of students has made the strongest bridge. If more than one bridge remains intact after the initial test, keep adding quarters until only one bridge remains standing.

Summary

Drawing a net and folding the sides together is one way to make a model of a three-dimensional shape. Another way is to make a bar model. The bars are the edges of the shape. The point at which two or more bars meet is called a vertex.

The key to making a structure more stable is to create more triangles within it.

Sometimes, the new triangles are parts of the original faces.

At other times, they are not.

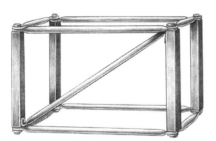
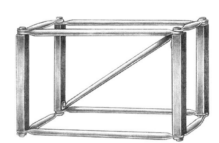

Summary Questions

Maha has six bars: three are 6 centimeters long, and three are 4 centimeters long.

9. She makes a triangle with one of the long bars and two of the shorter ones.

   a. As carefully and accurately as you can, draw this triangle.

   b. Draw a triangle with sides 4 centimeters, 6 centimeters, and 6 centimeters long.

10. Maha tries to make a prism with her six bars, but finds it impossible.

   a. Why can't Maha make a prism using just six bars?

   b. How many more bars does she need?

   c. What are possible lengths for the additional bars? Include a drawing to explain your answer.

   d. Would the prism you drew for part **c** be stable? Explain.

9. a.

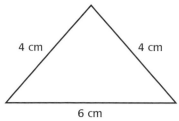

b.

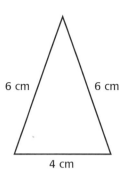

10. a. A prism has a minimum of nine edges; Maha has only six.

b. Maha needs three more bars to make a prism with a triangular base. If she makes a prism with a rectangular base, she will need at least six more bars.

c. Answers will vary. Sample answers and drawings:

i: three more 6-cm bars
ii: three more 4-cm bars
iii: one more 4-cm bar and two more 6-cm bars
iv: one more 6-cm bar and two more 4-cm bars

i.   **ii.**

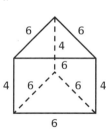   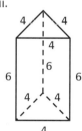

iii.   **iv.**

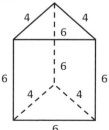   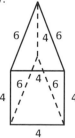

d. Answers may vary. Any of these constructions would be relatively stable but would shift horizontally because the sides are rectangles.

Materials centimeter rulers (one per student); gumdrops or modeling clay, optional (20 pieces per student); toothpicks, straws, or pipe cleaners, optional (6 cm and 4 cm long, 30 of each per student)

Overview Students read and discuss the Summary. They draw triangles and prisms using bars of two different lengths.

Planning Students may work on problems **9** and **10** individually. Problem **9** can be assigned as homework. Problem **10** can be used as an informal assessment. Discuss students' answers. After students complete Section C, you may assign appropriate activities from the Try This! section, located on pages 40–43 of the *Packages and Polygons* Student Book, for homework.

Comments about the Problems

9. Homework This problem can be assigned as homework. Students might make the shapes first, and then make drawings.

10. Informal Assessment This problem assesses students' ability to recognize and identify properties of regular polygons and polyhedra. It also assesses their ability to construct geometric models, draw two- and three-dimensional figures, and develop spatial visualization skills.

Some students may be able to answer the questions and make the drawings without having to construct the prisms. You may want to have bars available for students who need them.

This problem builds on problem **9,** in which students constructed triangles that could be used as faces of the prism. The drawings in the Solutions and Samples column on this page are of prisms with a triangular base, but the questions could be answered for a rectangular prism as well.

c. As a class, make a list of possible solutions.

Extension As a puzzle, ask students to use six sticks of 4 centimeters to form four equilateral triangles. There are two possible solutions, as shown below, although students are not expected to come up with the two-dimensional solution.

Two-dimensional

Three-dimensional (pyramid)

SECTION D. POLYGONS

Work Students Do

Students explore the characteristics of polygons by investigating the number of ways a shape fits onto itself. They learn what a polygon is and identify several common polygons. They compare regular and irregular polygons, considering what makes some polygons regular. Students construct polygons by moving from point to point on a clock. Then they use the concept of turns, or exterior angles, to find the relationship between a polygon and the sum of its interior angles. Finally, students find the radii of several circles that have been drawn around polygons.

Goals

Students will:

- recognize and identify geometric shapes and structures in real objects and in representations;*
- use the relationships between angles and turns to solve problems;
- recognize and identify properties of regular polygons and polyhedra;
- construct geometric models;*
- draw two- and three-dimensional figures.

These goals are assessed in other sections of the unit.

Pacing

- approximately three 45-minute class sessions

Vocabulary

- polygon
- regular polygon

About the Mathematics

The faces of a polyhedron are polygons. By exploring turns and thinking about how a shape fits onto itself, students discover that regular polygons have equal angles and equal sides and so can "fit" onto themselves in more ways than non-regular polygons. If you have an *n*-sided regular polygon, it can fit onto itself 2*n* ways (allowing flips). Turns can perhaps be best understood by thinking about walking around the edge of a polygon. When you arrive back at the point of origin, you have turned 360°.

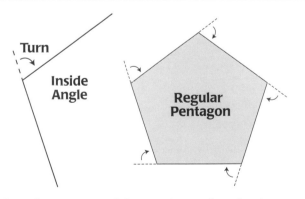

This exploration is used to introduce formulas for finding the measure of the interior angles of polygons. When the polygon is regular, dividing 360 by the number of turns determines the size of the exterior angles (another word for turns). The sum of the measures of each interior angle and exterior angle is 180°, so each interior angle can be found by subtracting the size of the exterior angle from 180°.

Materials

- Student Activity Sheets 1 and 2, pages 111 and 112 of the Teacher Guide (one of each per student)
- rectangular-shaped boxes, page 41 of the Teacher Guide, optional (one per group of students)
- cube-shaped boxes, page 41 of the Teacher Guide, optional (one per group of students)
- scissors, page 43 of the Teacher Guide (one pair per student)
- centimeter rulers, pages 45, 47, 49, and 55 of the Teacher Guide (one per student)
- compass cards or protractors, pages 45, 51, 53, and 55 of the Teacher Guide (one per student)
- tracing paper, pages 45 and 55 of the Teacher Guide (one sheet per student)
- different-colored pencils, page 49 of the Teacher Guide (three colors per student)

Planning Instruction

You may want to begin this section by providing students with some three-dimensional containers that have different kinds of covers. Then, have them experiment to see how many ways they can put the covers on the containers.

Students may work on problems 1–3 in small groups. Students may work on problems 4, 14, and 15 individually. The remaining problems may be done individually or in small groups.

There are no optional problems in this section.

Homework

Problems 4 (page 40 of the Teacher Guide), 7 and 8 (page 42 of the Teacher Guide), 14 and 15 (page 48 of the Teacher Guide), and 18 (page 52 of the Teacher Guide) can be assigned as homework. Also, the Extensions (pages 45 and 55 of the Teacher Guide) can be assigned as homework. After students complete Section D, you may assign appropriate activities from the Try This! section, located on pages 40–43 of the *Packages and Polygons* Student Book. The Try This! activities reinforce the key mathematical concepts introduced in this section.

Planning Assessment

- Problem 10 may be used to informally assess students' ability to recognize and identify the properties of regular polygons and polyhedra and to draw two- and three-dimensional figures.
- Problem 17 may be used to informally assess students' ability to use the relationships between angles and turns to solve problems.

D. POLYGONS

Put a Lid on It

Susanne has a construction kit with pieces of various shapes. She can use the pieces to build different boxes.

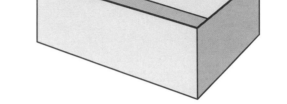

First, she builds a rectangular box. In the picture, you can see that the box is almost complete; only one more piece is needed.

1. In how many ways can she fit the last piece on top?

Next, Susanne wants to make a cube.

2. a. What shapes does she need? How many pieces of each shape does she need?

   b. In how many ways can she fit the last piece on top?

On the right, you see the third solid Susanne is making with her construction kit.

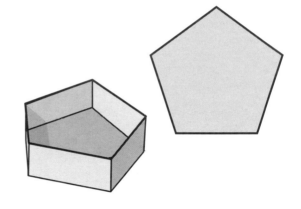

3. a. In how many ways can she fit the last piece on top?

   b. What is the name for the shape of this last piece?

The pieces in Susanne's construction kit are called polygons.

4. Shown below are some polygons from Susanne's kit. Study the shapes. Look for some similarities and write your own definition of a polygon.

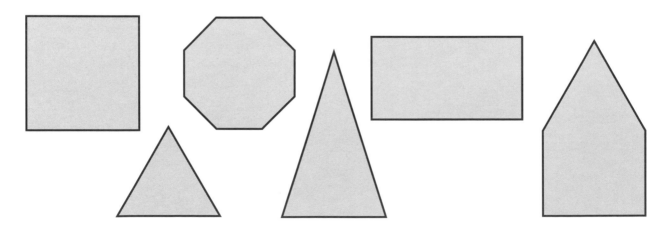

1. The last piece will fit on top in four different ways.

Top view (one angle marked with a dot):

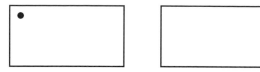

Flipping the piece over will give the other two possibilities.

2. a. She will need six square pieces to make a cube.

   b. The last piece will fit on top in eight different ways. Four possibilities:

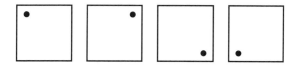

   Flipping the piece over will give the other four possibilities.

3. a. The last piece will fit on the top in 10 different ways.

   b. This shape is called a pentagon.

4. Answers will vary. Sample student response:

A polygon is a two-dimensional, closed figure with at least three sides and three or more corners.

Materials rectangular boxes, optional (one per group of students); cube-shaped boxes, optional (one per group of students)

Overview Students investigate the different ways in which a cover can be put on containers of different shapes. They write a definition of a polygon in their own words.

About the Mathematics Polygons are introduced here. The covers of the cube-shaped box and the pentagon-shaped box are regular polygons. Without flipping, the cover will fit as many times as the number of angles in the shape, since in a regular polygon all angles are the same and all sides are equally long.

Planning Students may work on problems **1–3** in small groups. Discuss their answers. Students may work on problem **4** individually. This problem can also be assigned as homework. Discuss students' definitions in class.

Comments about the Problems

   1. After having found two ways to place the cover, students may not think to flip the cover over. You may want to suggest that the cover has two sides and that it can be flipped upside down.

   2. If students are having difficulty, you might have them number the angles at the top of the cube. Then have them letter one angle on each side of the cover. Students can keep track of how many ways there are to fit the cover on the cube by recording the different number and letter combinations.

   3. b. Students may or may not be familiar with the word *pentagon*. Ask them to think of other places in which the root word *penta* is used, such as the pentathlon, an athletic competition made up of five events, or the Pentagon, a five-sided building near Washington, D.C., which is the headquarters for the U.S. Department of Defense.

   4. Homework This problem may be assigned as homework. The word *polygon* comes from the Greek word *polygonos*, which means "many angled." It is important to give students the opportunity to formulate their own definitions. Students should consider that the sides are straight line segments and that each shape has three or more sides and angles.

A polygon with three sides can be called a "3-gon," but it is usually called a triangle.

A polygon with four sides can be called a "4-gon."

5. a. What word is more commonly used for a polygon with four sides?

   b. What does the shape called a pentagon look like?

Susanne makes four prisms. Each prism is missing its lid.

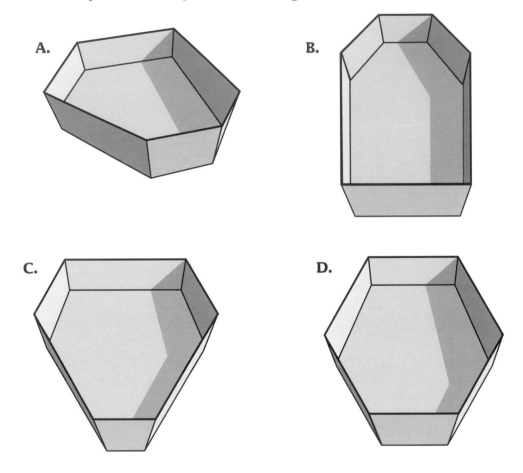

A. **B.**

C. **D.**

Use **Student Activity Sheet 1** to cut out each hexagonal lid.

6. In how many ways does hexagon A fit on top of prism A on this page? hexagon B on top of prism B? In how many ways do hexagons C and D fit on top of their prisms?

7. Compare the four hexagonal lids. How are their shapes similar? How do they differ?

Hexagon D is a special type of polygon; it is a regular polygon.

8. What do you think makes hexagon D "regular"?

5. a. A polygon with four sides is called a quadrilateral. If all the sides and angles are equal, a quadrilateral is called a square.

  b. A pentagon is a shape with five sides and five angles.

6. Hexagon A will fit on prism A one way.

  Hexagon B will fit on prism B two ways.

  Hexagon C will fit on prism C six ways.

  Hexagon D will fit on prism D 12 ways.

7. Answers will vary. Sample response:

  The shapes are similar because they all have six sides and six angles. The shapes are different because the lengths of their sides are different.

8. Answers will vary. Sample response:

  What makes hexagon D regular is all that the sides and angles are equal. If you fit this hexagon on prism D, it does not matter how you rotate the cover because all angles are equal and all sides are equally long. The cover will fit on prism D in six different ways without being flipped over.

Materials Student Activity Sheet 1 (one per student); scissors (one pair per student)

Overview Students learn the terminology for 3-, 4-, 5-, and 6-gons. They study properties of regular and irregular hexagons.

Planning Students may work on problems **5–8** individually or in small groups. Problems **7** and **8** can also be assigned as homework. If so, students should take the cutouts from Student Activity Sheet 1 home.

Comments about the Problems

5. Discuss the terminology for polygons with the class. Note that *gon* is not used in the terms *triangle, quadrilateral,* or *square,* but it is used in the terms *pentagon, hexagon, heptagon, octagon, nonagon, decagon,* etc. These latter terms are based on the Greek numbers.

6. Remind students that the prisms are three-dimensional, like boxes. The hexagons on Student Activity Sheet 1 are two-dimensional. They are the covers for prisms A–D.

7. Homework This problem may be assigned as homework. Students should focus on the sides and angles of the lids.

8. Homework This problem may be assigned as homework. Some students may have noticed regularity when they solved problem **7.**

Extension You may want to have students think of other words that use the prefixes *tri* and *quad,* such as *tricycle, tripod, quadruplets,* and *quadrangle.*

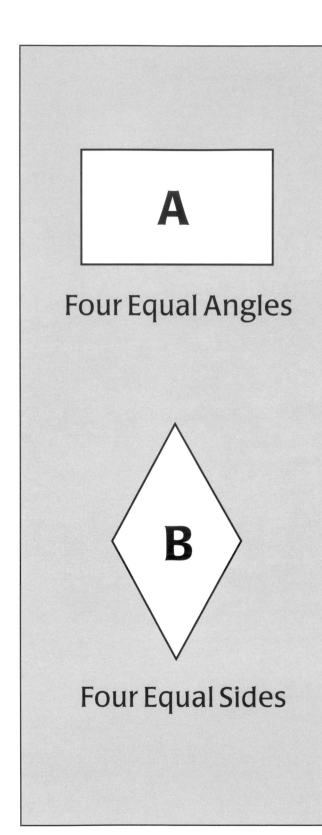

A

Four Equal Angles

B

Four Equal Sides

9. Quadrilaterals A and B, pictured on the left, are not regular polygons.

   a. What is irregular about figure A?

   b. What is irregular about figure B?

   c. In how many ways can figure A be folded in half so that it fits onto itself? figure B?

10. **a.** Draw a quadrilateral that can be folded in half four different ways.

   b. What is the name of this quadrilateral?

   c. Is the quadrilateral you drew in part **a** regular? Why or why not?

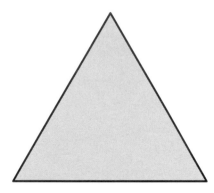

11. A regular triangle has three equal sides and three equal angles. The triangle shown above is regular.

   a. What is the measure of each angle?

   b. Draw a regular triangle with sides 8 centimeters long.

9. a. The sides of figure A have different lengths.

b. The angles of figure B have different measures.

c. Figures A and B can each be folded in half two ways, as shown below.

A

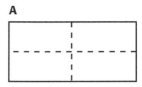

B

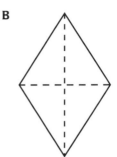

10. a. Students should draw squares as shown below.

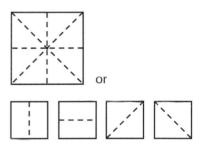

or

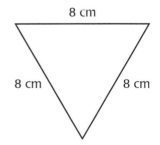

b. This quadrilateral is called a square.

c. It is regular, because the sides are all equal and so are the angles.

11. a. The measurement of each angle is 60°.

b. Drawings will vary. Sample drawing:

8 cm

8 cm   8 cm

Materials centimeter rulers (one per student); compass cards or protractors (one per student); tracing paper (one sheet per student)

Overview Students determine whether or not polygons are irregular by folding them. Students also investigate regular triangles.

About the Mathematics A polygon is regular if the number of possible half-folds is exactly equal to the number of sides. Students may already know that the angles of a triangle total 180° from the unit *Figuring All the Angles,* in which they learned to measure angles using a compass card. In that unit they also used this skill to measure angles in regular polygons.

Planning Students may work on problems **9–11** individually or in small groups. Problem **10** may be used as an informal assessment.

Comments about the Problems

9. Students can use a compass card or a protractor to measure angles.

   c. You may want to suggest that students trace figures A and B, cut them out, and fold them in order to solve this problem. *Note:* Each crease counts as only one fold, even though the shape could be folded forward and backward.

10. Informal Assessment This problem assesses students' ability to recognize and identify properties of regular polygons and polyhedra and to draw two- and three-dimensional figures.

Students should begin to recognize that regular polygons must have equal angles and equal sides.

11. b. One way to draw this triangle is to begin with a line 8 centimeters long. Then fold the paper so that the two ends of the line meet. Unfold the paper and note that the crease is a perpendicular line bisecting the first line through its center. The remaining two sides of the triangle must meet at one point along the perpendicular bisector.

Extension Have students try to find a relationship between the number of sides and the number of ways a regular polygon can be folded in half.

This is a photograph of the Pentagon building in Arlington, Virginia (outside of Washington, D.C.). The Pentagon is the headquarters of the U.S. Department of Defense and is one of the world's largest office buildings. The perimeter of the building is about 4,605 feet, and there are 17.5 miles of hallways. About 23,000 people work in the building.

12. Why do you think this building is named the Pentagon?

On the right, a drawing of the top view of the building has been started.

13. Trace this drawing and complete it to show a top view of the entire building.

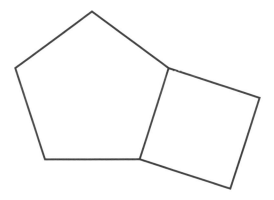

12. The Pentagon was named after a pentagon because it has five equal angles and five equal sides.

13.

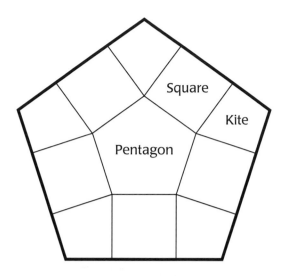

Materials centimeter rulers (one per student)

Overview Students study a photograph of the Pentagon. They explain how the Pentagon got its name and complete a top-view drawing of the entire structure.

Planning Students may work on problems **12** and **13** individually or in small groups.

Comments about the Problems

12. As they look at the photograph, students may conclude that all of the building's angles are equal and so are all of its sides. The word *pentagon* comes from a Greek word meaning five-angled.

13. In addition to the five squares in the sample solution to this problem, there are five other quadrilaterals that have two pairs of adjacent sides equal. These quadrilaterals are called *kites*.

Extension You can demonstrate making a pentagon by tying a knot in a strip of paper or a belt as shown below.

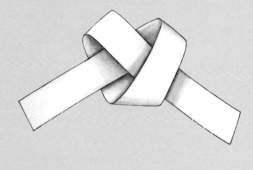

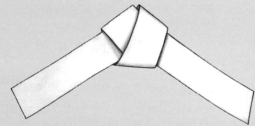

POLYGONS IN A CLOCK

You can use a picture of a clock to create regular polygons. Starting at one o'clock, make jumps of three hours. After four jumps, you are back to where you started. The result is a regular quadrilateral inside the clock.

14. Use **Student Activity Sheet 2** and a straightedge to make the following drawings. Be sure to count carefully.

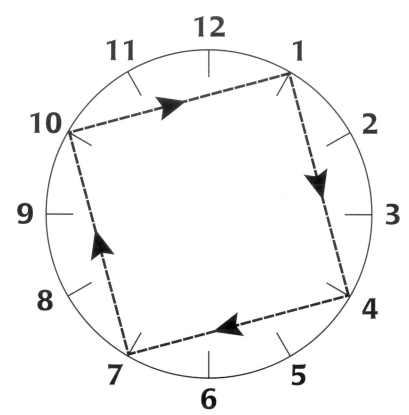

a. Start at one o'clock and make jumps of two hours until you are back at one o'clock. What polygon is formed?

b. Do the same thing as in part **a** with jumps of four hours. What is the result?

c. What polygon is formed if you make jumps of one hour?

d. Starting at 12 o'clock, make jumps of five hours. Continue making jumps of five hours until you are back at 12 o'clock. If you do this correctly, you will have a star. How many points does the star have?

15. Look again at the three-hour polygon shown above.

a. If you start at a whole number, in how many different positions on the clock can you place this polygon?

b. Draw all of these positions on one clock using different colors.

14. a. A hexagon is formed as shown below.

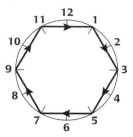

b. A triangle is formed.

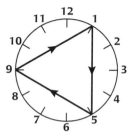

c. A dodecagon is formed as shown below.

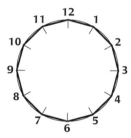

d. The star has 12 points, as shown below.

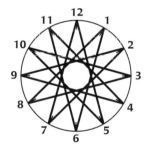

15. a. You can place this polygon at three different positions on the clock. You can start at one o'clock, two o'clock, or three o'clock. If you start at four o'clock, the square occupies the same space it did when you started at one o'clock.

b.

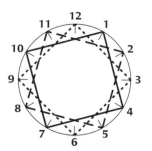

Materials Student Activity Sheet 2 (one per student); centimeter rulers (one per student); different-colored pencils (three colors per student)

Overview Students draw regular polygons on the face of a clock.

About the Mathematics An *n*-sided regular polygon can fit on itself *n* ways (without flipping).

Planning Students may work on problems **14** and **15** individually. These problems also can be assigned as homework. You may want to display students' drawings in the classroom or on the school bulletin board.

Comments about the Problems

14–15. These problems may be assigned as homework.

14. Students should use a ruler to make the drawings.

c. The word *dodecagon* comes from the Greek word *dodeca,* which means 12.

15. a–b. Some students may think that the polygon can be placed in 12 different positions. When students draw the positions, they will see that there are only three different ones.

You can find the measures of the angles of a polygon by using turns. To do this, imagine yourself walking along the edges of a polygon. Picture the angle that is made each time you turn a corner.

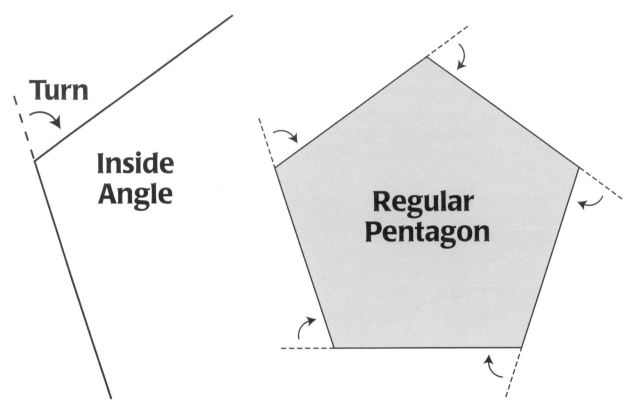

16. a. How many degrees would you turn if you walked all the way around a regular pentagon? a square? a regular triangle?

b. If you walked all the way around any polygon, how many degrees would you turn? Why?

17. a. When you walk around a regular pentagon, you make five equal turns. How many degrees are in each turn?

b. How many degrees are in each inside angle of a regular pentagon?

c. How can you use turns to find the measures of the inside angles of any regular polygon?

16. **a.** In each case you would turn 360°.

   b. If you walk all the way around any polygon, you would turn 360°. If you start walking, your face is pointed in a certain direction. You walk and turn, and after the last turn your face is in the same direction as when you started. So you turned one complete circle, which is 360°.

17. **a.** 72° (360° ÷ 5)

   b. 108°. The angle of a straight line (180°) minus the angle of the turn (72°) equals 108°. Some students may find this by drawing the angle:

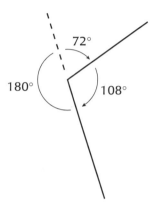

   c. Divide 360° by the number of turns. The answer is the measure of the angle of one turn. Then subtract this measure from 180° to get the measure of the inside angle.

Materials compass cards or protractors (one per student)

Overview Students discover how they can use the number of turns to find the measures of the inside angles of any regular polygon.

About the Mathematics Students learned about the relationships between turns and inside or resulting angles in the grade 5/6 unit *Figuring All the Angles*. If you make a turn, the turn and the resulting angle add up to 180°. In that unit, students learned about turns and angles from a motion point of view: they put themselves in a starting position and imagined walking around the figure. Turns and angles are described from their own perspective. You may have to remind students of this approach to turns and angles.

Planning Students may work on problems **16** and **17** individually or in small groups. Problem **17** may be used as an informal assessment.

Comments about the Problems

16. **a.** Some students may want to measure the turn using a compass card, as introduced in the unit *Figuring All the Angles*. If students measure, however, their answers may be a little off. Other students may be able to use reasoning to determine how big the turn is.

   b. This question asks students to generalize about what they learned in problem **16a.**

17. **Informal Assessment** This problem assesses students' ability to use the relationships between angles and turns to solve problems.

This is a picture of bees in their honeycomb.

18. a. What regular polygon do you see in this picture?

b. What is the measure of one inside angle of this polygon?

Summary

Polygons are two-dimensional shapes with three or more angles. Polygons are named according to the number of sides or angles they have. For instance, a polygon with three sides is called a triangle, a polygon with four angles is called a *quadrilateral*, and so on. Others are called *n*-gons, or polygons that have *n* sides. For example, a 9-gon is a polygon with nine sides.

A *regular polygon* has equal sides and equal angles. Knowing how many turns you would have to make to move around a regular polygon, you can find the measure for each inside angle. Just remember that to completely move around a regular polygon, you turn a total of 360°.

18. a. This picture contains hexagons.

b. The measure of one inside angle of this polygon is 120°. Strategies will vary. Sample strategies:

- Using turns:
  360° ÷ 6 = 60°
  180° − 60° = 120°

- Using regular triangles:

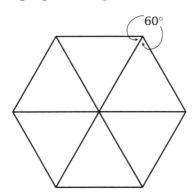

The inside angles of regular triangles are equal. Each angle measures 60°. Each honeycomb angle contains two triangle angles;
60° + 60° = 120°.

- Using a honeycomb pattern:

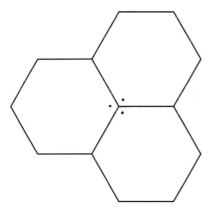

The angles designated with dots are equal, so 360° ÷ 3 = 120° per angle.

Materials compass cards, optional (one per student)

Overview Students find the measurement of each inside angle of a hexagon. They also read the Summary, which reviews the main concepts covered in this section.

About the Mathematics For regular polygons, the size of the turn (or exterior angle) can be found by dividing 360° by the number of turns. The sum of the size of a turn and the resulting (or interior) angle is 180°. Each interior angle can be found by subtracting the size of the turn from 180°.

Planning Students may work on problem **18** individually or in small groups. Problem **18** can also be assigned as homework.

Comments about the Problems

18. Homework This problem may be assigned as homework. In order to solve this problem, students may use the strategy they developed in problem **17,** using turns. However, other strategies are also possible. If students use triangles, remind them that the angles of a triangle add up to 180°. If students use the honeycomb pattern, they should see that, together, the three angles that meet measure 360°. Measuring with a compass card is not necessary to solve this problem using any of the sample strategies shown in the solutions column. Encourage students to solve the problem using reasoning. Also encourage them to make any sketches they think will help them.

Summary Questions

Shown below are seven regular polygons, each with sides of 4 centimeters. Each polygon can be surrounded by a circle, as shown below to the left with the triangle. Each circle's center point is *M*, and each circle intersects the vertices of the polygon.

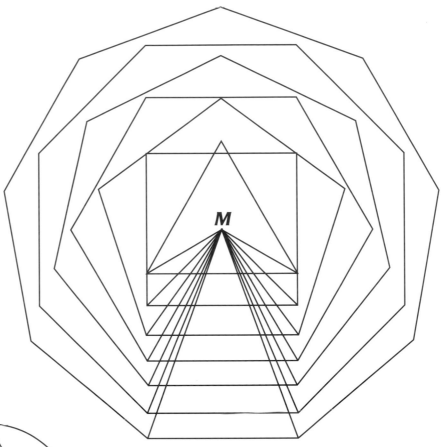

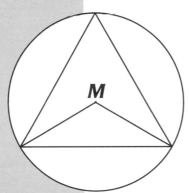

19. a. The circle in which the triangle fits is the smallest one. What is the length of this circle's radius?

b. How long would the radius of the biggest circle be, the one in which the 9-gon would fit?

20. From the center, *M*, two lines are drawn to neighboring vertices of each regular polygon.

a. Measure the angle between the two lines drawn for the 9-gon.

b. How can you determine the number of degrees in this angle without measuring it?

19. a. The length of the circle's radius is 2.3 centimeters.

   b. The length of the circle's radius would be 5.9 centimeters.

20. a. The angle measures about 40°.

   b. 360° ÷ 9 = 40°. Some students may find this answer by drawing lines to each vertex:

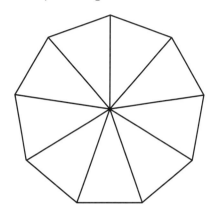

If lines are drawn to each vertex of the 9-gon, nine identical triangles are created. The angles in the center of the 9-gon must add up to 360°. Each angle must be $\frac{1}{9}$ of 360°, or 360° ÷ 9, which is 40°.

Materials centimeter rulers (one per student), compass cards or protractors (one per student), tracing paper (one sheet per student)

Overview Students answer a series of questions involving a set of seven regular polygons. Each polygon has equal sides and the same center point. Students determine the lengths of the radiuses of two circles that can be drawn around two of the polygons. Students also find the measurements of angles by measuring and reasoning.

About the Mathematics At this stage, students will not be able to calculate the radius of a circle; they can only measure it.

Planning Students may work on problems **19** and **20** individually or in small groups. Discuss their answers. After students complete Section D, you may assign appropriate activities from the Try This! section, located on pages 40–43 of the *Packages and Polygons* Student Book, for homework.

Comments about the Problems

19. Students can find the answers to **19a** and **b** by measuring.

20. a. Students may measure the angle using a compass card, as they learned in the grade 5/6 unit *Figuring All the Angles.*

   b. Note that students can find the precise measurement of the angle by reasoning. If they measure the angle with a compass card, the measurement may be a little bit off.

Extension You may want students to investigate this figure more extensively. They can measure all of the sides and find all of the angles. Students can also organize their results in a chart. Ask students, *What would a regular 100-gon look like? How would you estimate the radius of a regular 100-gon with a side of 4 centimeters?* [Students should not need to draw the 100-gon. After measuring the length of the radius of each circle that could be drawn around the polygons shown here, students should note that the radius grows almost linearly, about 0.8 centimeters each time a side is added to the polygon. Since 0.8 × 100 = 80, the answer is approximately 80 centimeters.]

SECTION E. PLATONIC SOLIDS

Work Students Do

Students identify and name the five regular polyhedra. They make solid models of the polyhedra using nets made from posterboard and look for patterns based on the relationships between the numbers of faces, edges, and vertices. Then, they consider criteria for the regularity of polyhedra. Students learn that they can use a cube as a package for a tetrahedron. They place a tetrahedron in a cube and color faces of the tetrahedron in a two-dimensional drawing. Students complete a table of data about the five Platonic solids.

Goals

Students will:

- recognize and identify properties of regular polygons and polyhedra;*
- construct geometric models;
- make connections between different views of geometric solids;*
- develop efficient counting strategies, involving geometric solids, that can be generalized;
- reason about the structure of the Platonic solids;
- develop spatial visualization skills.

  These goals are assessed in other sections of the unit.

Pacing

- approximately two or three 45-minute class sessions

Vocabulary

- Platonic solids
- polyhedron
- regular polyhedron
- tetra-

About the Mathematics

The five regular polyhedra (tetrahedron, octahedron, icosahedron, cube, and dodecahedron) are the focus of this section. The notion of "regular" is introduced again using the context of tracing a face on the ground and seeing how many ways the regular polyhedron will fit onto it. Students are not expected to give a formal definition, but they should be able to see that all the faces of a regular polyhedron are congruent regular polygons, and that the same number of edges intersect at each of its vertices. Counting and reasoning about vertices, edges, and faces can help students develop efficient counting techniques.

Materials

- Student Activity Sheets 3–9, pages 113–119 of the Teacher Guide (one of each per student)
- posterboard, page 61 of the Teacher Guide (five sheets per student or group of students)
- scissors, page 61 of the Teacher Guide (one pair per student)
- tape, pages 61 and 67 of the Teacher Guide (one dispenser per group of students)
- pipe cleaners or string, page 61 of the Teacher Guide (90 segments per group of students)
- straws, page 61 of the Teacher Guide (90 per group of students)
- models of polyhedra made by students on page 24 of the Student Book, page 60 of the Teacher Guide, pages 63, 65, 67, and 69 of the Teacher Guide, optional (one of each per group of students)
- colored pencils or crayons, page 67 of the Teacher Guide (one box per student)
- blank transparency, page 67 of the Teacher Guide, optional (one per class)

Planning Instruction

This section introduces students to the regular polyhedra. You may want to begin the section by discussing the polyhedra pictured on page 58 of the Teacher Guide. Ask students, *What polygons are used as faces for the special polyhedra?* [triangles, squares, and pentagons]

Students may work on problems 6 and 8 in small groups. They may do problems 9–11 individually. The remaining problems may be done individually or in small groups.

There are no optional problems in this section.

Homework

Problems 3 (page 60 of the Teacher Guide) and 7 (page 64 of the Teacher Guide) can be assigned as homework. Also, the Extension (page 63 of the Teacher Guide) can be assigned as homework. After students complete Section E, you may assign appropriate activities from the Try This! section, located on pages 40–43 of the *Packages and Polygons* Student Book. The Try This! activities reinforce the key mathematical concepts introduced in this section.

Planning Assessment

- Problem 5 may be used to informally assess students' ability to construct geometric models.
- Problem 9 may be used to informally assess students' ability to develop spatial visualization skills.
- Problems 10 and 11b may be used to informally assess students' ability to reason about the structure of the Platonic solids.
- Problem 11a may be used to informally assess students' ability to develop efficient counting strategies, involving geometric solids, that can be generalized.

E. PLATONIC SOLIDS

Special Polyhedra

A *polyhedron* is a three-dimensional shape whose faces are all polygons. The word polyhedron comes from the Greek words for "many bases." You have already worked with many different polyhedra.

1. Look again at the pictures on pages 2 and 3. Which of the shapes are polyhedra?

Here are five special polyhedra.

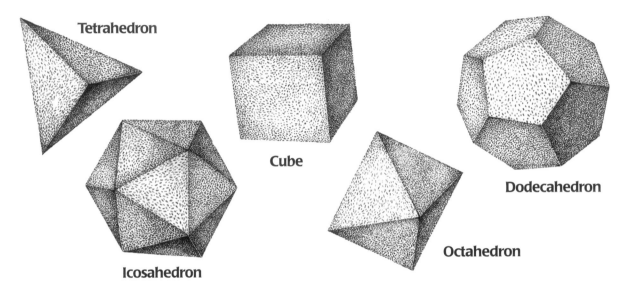

Tetrahedron

Cube

Dodecahedron

Octahedron

Icosahedron

Below are bar models for the five special polyhedra.

2. What is each bar model's name?

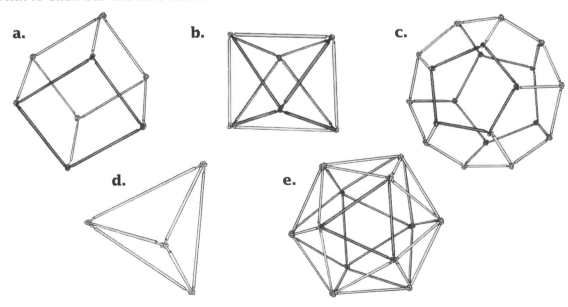

a.

b.

c.

d.

e.

1. The following shapes are polyhedra:

   prism

   pyramid

   cube

   rectangular prism

2. **a.** cube

   b. octahedron

   c. dodecahedron

   d. tetrahedron

   e. icosahedron

Overview Students learn to recognize polyhedra and to name bar models of polyhedra.

Planning Students may work on problems **1** and **2** individually or in small groups.

Comments about the Problems

1. Students should note that each shape has faces that are identical, regular polygons. They may also see that the same number of edges intersect at each vertex.

2. The names for the five special polyhedra are shown with the solid models above problem **2**. Students may count the faces to name the polyhedra:

   cube: six faces
   octahedron: eight faces
   dodecahedron: twelve faces
   tetrahedron: four faces
   icosahedron: twenty faces

Did You Know? Except for the cube, the regular polyhedra all have names based on Greek numbers and the Greek word for base, *hedron.* Explain to students that the plural of *hedron* is *hedra* and the plural of *polyhedron* is *polyhedra.*

Activity

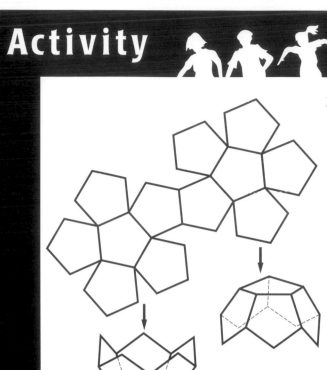

3. Trace the nets on **Student Activity Sheets 3–7** onto heavy paper and cut them out to make the five regular polyhedra.

The drawings on the left illustrate how you can make the regular dodecahedron from the net. The dodecahedron is made from two baskets that fit together.

Keep your models handy. They will help you answer questions throughout this unit.

Look at the bar models on page 23. You can make these models with drinking straws. You can use pipe cleaners or string to make connections at the vertices.

Three edges meeting at one vertex

4. To make all five models, you need a total of 90 straws. How many straws does each bar model require?

5. You have already made model a. Now make models b, c, d, and e.

3. Students' posterboard models should look like the ones pictured on page 58 of the Teacher Guide.

4. Models **a** and **b** require 12 straws.
Models **c** and **e** require 30 straws.
Model **d** requires 6 straws.
The total number of straws required is 90.
Students may count the straws that were used in the bar models shown on page 58 of the Teacher Guide.

5. Students' bar models should look like the bar models pictured on page 58 of the Teacher Guide.

Materials Student Activity Sheets 3–7 (one of each per student); posterboard (five sheets per student or group of students); scissors (one pair per student); tape (one dispenser per group); pipe cleaners or string (90 segments per group); straws (90 per group)

Overview Students construct paper models of the five regular polyhedra using nets, and bar models using straws. They determine how many straws (edges) they need to build each polyhedron.

Planning Students may work on problems **3–5** individually or in small groups. If time is a concern, problem **3** can also be assigned as homework. Problem **5** may be used as an informal assessment. To save time and material costs, you may prefer to have students work in small groups on these problems. Each group could make one of each model.

Comments about the Problems

3–5. Have students save the finished models. They will need to use them to answer questions throughout this section.

3. Homework This problem may be assigned as homework. Encourage students to count vertices, faces, and edges. Discuss the characteristics that all regular polyhedra share:
- If the polyhedron's faces are squares or pentagons, exactly three faces must meet at each vertex.
- If the polyhedron's faces are triangles, three, four, or five faces may meet at its vertices.
- Polygons with six or more sides cannot be arranged to form a regular polyhedron.

5. Informal Assessment This problem assesses students' ability to construct geometric models.

Extension If students have difficulty understanding why there are only five regular polyhedra, you might tell them the story of the flat polyhedron: One student was working very hard making a dodecahedron, but she had not looked very carefully at the picture and made hexagons instead of pentagons. She could not construct a three-dimensional solid. Her net is shown below. You may wish to have students try this themselves.

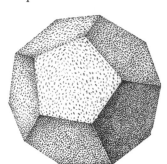

You just made paper and bar models of the five famous *Platonic solids*. We call them "Platonic" after the ancient Greek philosopher Plato.

These five solids are the only *regular polyhedra* that exist. It is not possible to construct other regular polyhedra.

All of the Platonic solids occur in nature as crystals. The microscopic diamond crystals shown below are in the shape of octahedra.

6. These students made a large bar model of a tetrahedron. One of them outlines the base with chalk.

   a. In how many ways can the tetrahedron fit on the chalk outline?

   b. In how many ways can an octahedron fit on an outline of one of its faces?

6. a. 12 different ways. There are four faces. Each face has three edges and can be placed on the ground three ways ($3 \times 4 = 12$).

b. 24 different ways. There are eight faces. Each face has three edges and can be placed on the ground three ways ($3 \times 8 = 24$).

Materials models of polyhedra made by students on page 24 of the Student Book, optional (one of each per group of students)

Overview Students investigate how many different ways a bar model of a regular polyhedron can fit on an outline of one of its faces.

About the Mathematics All the faces of a regular polyhedron are congruent regular polygons, and the same number of edges intersect at each of its vertices.

Planning Students may work on problem **6** in small groups. If students are having difficulty with this concept, discuss their answers in class.

Comments about the Problems

6. Some students will be able to find the answers by reasoning. Other students will need the models they constructed in problems **3** and **5.** They could draw an outline and develop a technique for counting the different ways the model can fit on the outline. For example, students could number the faces of the paper models.

Extension The following problems could be assigned as homework:

- How many different ways could the three students fit an icosahedron on a chalk triangle? [$3 \times 20 = 60$]

- Suppose that the three students drew chalk outlines of one face of a cube and one face of a dodecahedron. How many times would the cube and the dodecahedron be able to fit onto the outlines of their faces? [In the case of the cube, $4 \times 6 = 24$; for the dodecahedron, $5 \times 12 = 60$.] Students may want to number the faces to keep track of the combinations.

7. The octahedron on the left calculated the number of his edges, even though he cannot see them.

I HAVE 8 FACES; EACH FACE HAS 3 EDGES: SO I HAVE 24 EDGES!

a. Do you agree with the octahedron? Explain.

b. The cube seems to know better. In your notebook, complete his reasoning.

I HAVE 6 FACES; _____

_____

BUT, _____

_____!

SO I HAVE _____ EDGES!

30 EDGES, YES I DO!

c. The icosahedron says she has 30 edges. Is this true? How can you be sure?

7. a. No. The octahedron counted each edge twice. Because each edge belongs to exactly two faces, he needs to divide 24 by 2 to get the correct number of edges, 12.

b. Answers will vary. Sample response:

I have 6 faces; each face has 4 edges; $6 \times 4 = 24$ edges. But, now I have counted each edge twice, so I must divide by 2! So I have 12 edges!

c. Yes, the icosahedron is correct. She has 20 faces; each face has three edges. $20 \times 3 = 60$, but since each edge has been counted twice, you divide by two to find the actual number of edges; $60 \div 2 = 30$ edges.

Materials models of polyhedra made by students on page 24 of the Student Book, optional (one of each per group of students)

Overview Students are challenged to determine the number of edges of regular polyhedra by reasoning.

Planning Students may work on problem **7** individually or in small groups. This problem can also be assigned as homework. Discuss students' reasoning in class.

Comments about the Problems

7. Homework This problem may be assigned as homework.

   a. Some students may conclude that something is wrong with the octahedron's reasoning, because they can count only 12 edges on the octahedron.

   b. The cube has a different number of faces than the octahedron, but the fact that each edge belongs to two faces remains the same.

   c. Encourage students to reason in the same way as they did in problems **7a** and **7b.** Students may also check their answer by counting the edges of the posterboard or bar model.

Did You Know? Plato (429–348 B.C.) was a geometer (a specialist in geometry) as well as a philosopher. Plato was fascinated by the five regular polyhedra. Like many ancient Greeks, he believed that matter was made out of four elements: fire, water, air, and earth. Plato thought that the four elements were made of small particles shaped like regular polyhedra—fire had the form of a tetrahedron, water an icosahedron, air an octahedron, and earth a cube. Scientists today know that many substances are made up of atoms arranged in the shapes of regular polyhedra. The regular polyhedra came to be called *Platonic solids* because of the importance Plato placed on them as the building blocks of the universe.

The simplest Platonic solid is the tetrahedron. The prefix *tetra-* means "four." The tetrahedron has four faces (triangles), four vertices, and six edges.

You can use your cube as a package for the tetrahedron.

Open your cube at one side. The tetrahedron fits into the cube so that the cover can be closed again. Try it.

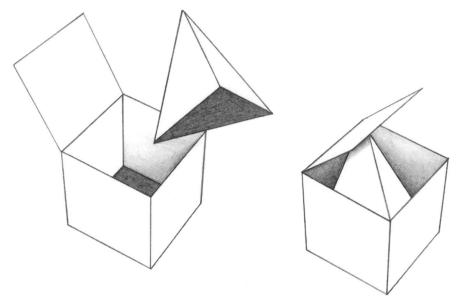

8. In how many different ways does the tetrahedron fit into the cube?

9. Shown below are pictures of a transparent cube with a tetrahedron inside. On **Student Activity Sheet 8,** color one face of the tetrahedron in each of the pictures.

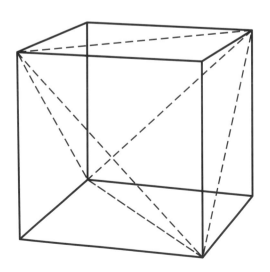 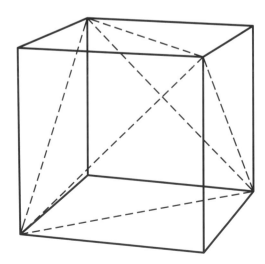

8. There are two or four upright positions for each edge; $2 \times 6 = 12$, and $4 \times 6 = 24$, so there are 12 or 24 possible ways. Two of them are shown below:

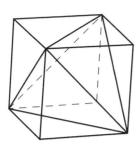 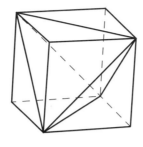

9. Answers will vary. Sample responses:

Left Cube	Right Cube

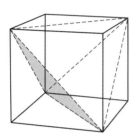 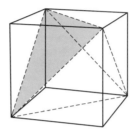

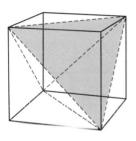 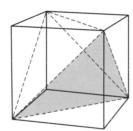

Materials Student Activity Sheet 8 (one per student); tape, optional (one dispenser per class); models of polyhedra made by students on page 24 of the Student Book, optional (one of each per group of students); colored pencils or crayons, (one box per group); blank transparency, optional (one per class)

Overview Students find out in how many ways the tetrahedron fits into the cube. They color faces of the tetrahedron in a two-dimensional drawing.

Planning Students may work on problem **8** in small groups. Discuss their answers. Problem **9** can be done individually and may be used as an informal assessment.

Comments about the Problems

8. Because all the faces look the same, some students may say that there are only four ways, two ways, or even only one way that the tetrahedron fits inside the cube. If students have difficulty, you might have them color each face of the tetrahedron a different color.

9. Informal Assessment This problem assesses students' ability to develop spatial visualization skills.

If students have difficulty visualizing the faces of the tetrahedron within the cube, you might make a transparent cube using a blank transparency sheet and tape. Use the net from Student Activity Sheet 4 and the tetrahedron made earlier with each face colored.

Summary

There are five Platonic solids. They are the only polyhedra that are regular. You can find many relationships among the edges, faces, and vertices of the Platonic solids.

Summary Questions

10. Explain why the Platonic solids are regular polyhedra.

11. a. Complete the table on **Student Activity Sheet 9.** To help you, use the five models you made.

   b. Study the numbers in your table. What patterns or relationships do you see?

Name	Shape	Type of Face	Number of Faces	Number of Vertices	Number of Edges
Tetrahedron		triangle	4	4	6
Cube					
Octahedron					
Dodecahedron					
Icosahedron					

10. For each Platonic solid, you can say that the faces are regular polygons and an equal number of edges must meet at each vertex. The faces of three Platonic solids are all regular triangles. The other two have faces that are all squares or faces that are all regular pentagons.

11. a.

Name	Shape	Type of Face	Number of Faces	Number of Vertices	Number of Edges
Tetrahedron			4	4	6
Cube			6	8	12
Octahedron			8	6	12
Dodecahedron			12	20	30
Icosahedron			20	12	30

b. Answers will vary. Some students may discover that the number of faces plus the number of vertices is always two more than the number of edges.

Materials Student Activity Sheet 9 (one per student); models of polyhedra made by students on page 24 of the Student Book, optional (one of each per group of students)

Overview Students read the Summary, which reviews the main concepts covered in this section. Students explain why Platonic solids are regular polyhedra. They also search for relationships between the numbers of faces, edges, and vertices of regular polyhedra.

About the Mathematics For regular polyhedra, the number of faces plus the number of vertices is always two more than the number of edges. This is Euler's formula: faces + vertices − edges = 2. This formula will be explored further in the next section.

Planning Students may work on problems **10** and **11** individually. These problems may also be used as informal assessments. Discuss students' answers. After students complete Section E, you may assign appropriate activities from the Try This! section, located on pages 40–43 of the *Packages and Polygons* Student Book, for homework.

Comments about the Problems

10, 11b. Informal Assessment These problems assess students' ability to reason about the structure of Platonic solids.

10. Students do not need to give formal definitions in response to this problem. They may use their own words.

11. a. Informal Assessment This problem assesses students' ability to develop efficient counting strategies, involving geometric solids, that can be generalized.

Some students may realize that they can find the number of edges at a vertex (for example, three for a tetrahedron). Then, find the product of the edges and the number of vertices (12 for a tetrahedron). Finally, divide by two because each edge was counted twice. (Therefore, a tetrahedron has six edges.)

11. b. Give students time to explore this table. Encourage them to explain the patterns they find in their own words.

SECTION F. EULER'S FORMULA

Work Students Do

In this section, students further explore the relationship among the numbers of faces (F), vertices (V), and edges (E) of other solids. They learn the formal relationship known as Euler's formula, $F + V - E = 2$. Students investigate the formula for regular polyhedra and other polyhedra. They are encouraged to think about strategies for counting the numbers of faces, vertices, and edges for a given polyhedron. Students check their work by using the results in Euler's formula.

Goals

Students will:

- understand and use Euler's formula;
- recognize and identify properties of regular polygons and polyhedra;*
- draw two- and three-dimensional figures;
- make connections between different views of geometric solids;
- develop efficient counting strategies, involving geometric solids, that can be generalized;
- reason about the structure of the Platonic solids*;
- develop spatial visualization skills;
- solve problems involving geometric solids.

  * These goals are assessed in other sections of the unit.

Pacing

- approximately one or two 45-minute class sessions

About the Mathematics

This section introduces Euler's formula, which expresses the relationship between the numbers of faces, vertices, and edges of solids. The problems in this section continue to build on students' ability to reason abstractly about the numbers of faces, vertices, and edges in a given polyhedron and to validate their reasoning either by counting or by using the values they find in Euler's formula to find other values.

Materials

- models of polyhedra made by students on page 24 of the Student Book, page 73 of the Teacher Guide, optional (one of each per student or group of students)
- straightedges or rulers, pages 75 and 79 of the Teacher Guide (one per student)
- models of a tetrahedron, page 77 of the Teacher Guide, optional (one per student)
- scissors, page 79 of the Teacher Guide, optional (one pair per student)

Additional Resources

- Burton, D. M., *The History of Mathematics: An Introduction* (Dubuque, IA: Wm. C. Brown Publishers, 1991)

Planning Instruction

You may want to introduce this section by discussing the life of Leonhard Euler (1707–1783), a famous mathematician. Euler (pronounced "oiler") began his career in Switzerland. He received his master's degree in mathematics at the age of 19. Then, Euler attended the Academy of St. Petersburg in Russia, where he astonished Russian mathematicians by computing in three days astronomical tables that would normally require several months' work. Euler wrote many books throughout his career and received many awards for his work in mathematics and physics.

Students may work on problem 1 individually or in small groups. The remaining problems may be done individually.

There are no optional problems in this section.

Homework

Problems 1 (page 72 of the Teacher Guide), 3 and 4 (page 74 of the Teacher Guide), and 5 (page 76 of the Teacher Guide) may be assigned as homework. After students complete Section F, you may assign appropriate activities from the Try This! section, located on pages 40–43 of the *Packages and Polygons* Student Book. The Try This! activities reinforce the key mathematical concepts introduced in this section.

Planning Assessment

- Problem 2 may be used to informally assess students' ability to draw two- and three-dimensional figures, and to make connections between different views of geometric solids.
- Problem 6 may be used to informally assess students' ability to understand and use Euler's formula, to draw two- and three-dimensional figures, and to develop spatial visualization skills.
- Problem 7 may be used to informally assess students' ability to understand and use Euler's formula; to develop efficient counting strategies, involving geometric solids, that can be generalized; and to solve problems involving geometric solids.

F. EULER'S FORMULA

Faces, Vertices, and Edges

In the previous section, you discovered a relationship between the numbers of faces, vertices, and edges of a polyhedron. This relationship is explained in Euler's formula:

$$\text{Number of Faces} + \text{Number of Vertices} - \text{Number of Edges} = 2$$

It is named for the Swiss mathematician Leonhard Euler. The relationship is usually written the following way:

$$F + V - E = 2$$

1. Does Euler's formula work for the polyhedra pictured below? Explain.

a.

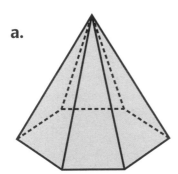

b.

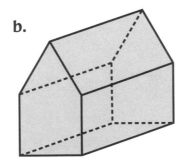

c.

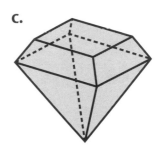

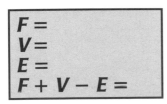

$F =$
$V =$
$E =$
$F + V - E =$

$F =$
$V =$
$E =$
$F + V - E =$

$F =$
$V =$
$E =$
$F + V - E =$

1. Yes, the formula works in each case.

   a. $F = 7$
      $V = 7$
      $E = 12$

      $F + V - E = ?$
      $7 + 7 - 12 = 2$

   b. $F = 7$
      $V = 10$
      $E = 15$

      $F + V - E = ?$
      $7 + 10 - 15 = 2$

   c. $F = 9$
      $V = 9$
      $E = 16$

      $F + V - E = ?$
      $9 + 9 - 16 = 2$

Materials models of polyhedra made by students on page 24 of the Student Book, optional (one of each per student or group of students)

Overview Students learn about Euler's formula. They check to see whether Euler's formula works for a number of pictured polyhedra.

Planning Students may work on problem **1** individually or in small groups. This problem can also be assigned as homework.

Comments about the Problems

  1. Homework This problem may be assigned as homework. It is critical to this section. Students need to be able to count the numbers of faces, edges, and vertices of different solids and determine whether the relationship given by Euler's formula holds true for these polyhedra. You may want to have the models students made of polyhedra in Section E available for them to use to verify Euler's formula.

Cutting and Polishing Diamonds

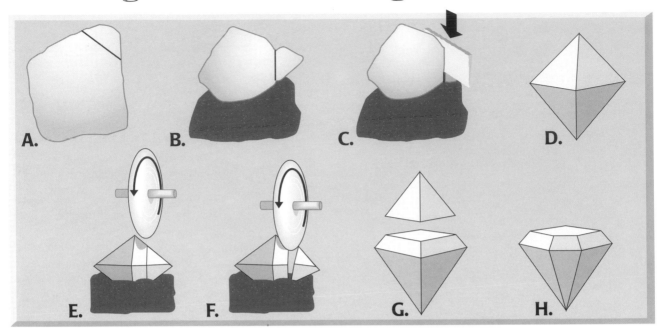

Look at the pictures above. Pay particular attention to the shapes in stages G and H. The larger shape in stage G has 9 faces, 9 vertices, and 16 edges.

2. Make a top-view drawing of the larger shape in stage G.

You get the shape in stage H by cutting away a corner from the larger shape in stage G.

3. a. Make a top-view drawing of the shape in stage H.

  b. How many faces does it have?

  c. How many vertices?

  d. How many edges?

  e. Calculate $F + V - E$ for the shape in stage H.

4. a. Draw a top view of how the shape in stage H looks after the other three corners have been cut away.

  b. How many faces, edges, and vertices does the diamond have now?

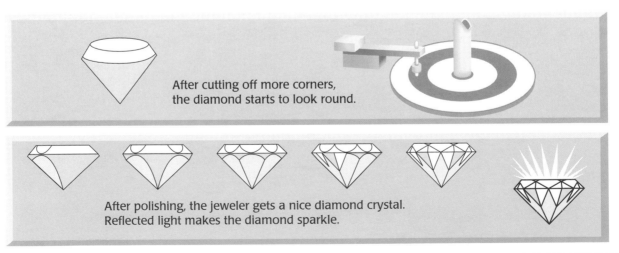

After cutting off more corners, the diamond starts to look round.

After polishing, the jeweler gets a nice diamond crystal. Reflected light makes the diamond sparkle.

2.

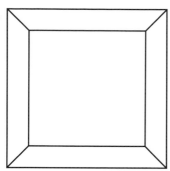

3. a.

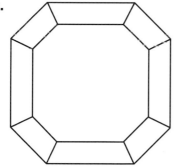

  b. $F = 11$

  c. $V = 11$

  d. $E = 20$

  e. $11 + 11 - 20 = 2$

4. a.

b. $F = 17$
    $V = 17$
    $E = 32$

Materials straightedges or rulers (one per student)

Overview Students learn about the cutting of diamonds and study pictures of the shapes at different stages. They check to see whether Euler's formula works for the shapes in two different stages.

Planning Students may work individually on problems **2–4.** You may wish to use problem **2** as an informal assessment. Problems **3** and **4** can be assigned as homework.

Comments about the Problems

  2. Informal Assessment This problem assesses students' ability to draw two- and three-dimensional figures and to make connections between different views of geometric solids.

     Students may remember making top views from the grade 5/6 unit *Side Seeing*. If students have difficulty, you may want to create a model of stage G for students to look at.

  3–4. Homework These problems may be assigned as homework.

  3. Note that in stage H, only one of the corners has been cut away.

  4. At this stage, some students may notice a pattern:

     Each time a corner is cut, the numbers of faces and vertices increase by 2 and the number of edges increases by 4, as shown in the table below.

	Original shape	1 corner cut	2 corners cut	3 corners cut	4 corners cut
F	9	11	13	15	17
V	9	11	13	15	17
E	16	20	24	28	32

Euler's formula, which you have explored with several different polyhedra, works for all polyhedra.

By cutting a diamond shaped like a tetrahedron, you can see why this formula works for all polyhedra.

The tetrahedron has four faces, four vertices, and six edges.

Euler's formula states that:

$$F + V - E = 2$$

Cutting off one tip of the tetrahedron, as in step 1 on the right, results in the following:

- F increases by 1

- V increases by 2

- E increases by 3

5. a. Why doesn't the formula $F + V - E$ change in step 1?

   b. Repeat the same process for steps 2, 3, and 4. Recalculate $F + V - E$ for each step.

The resulting figure after four steps is a semi-regular solid. The faces are regular triangles and regular hexagons.

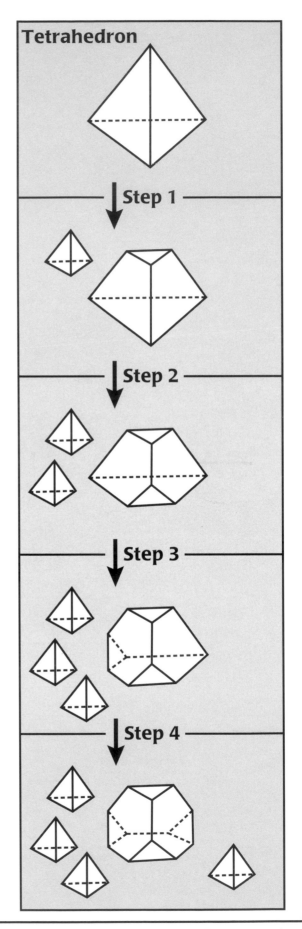

Tetrahedron

Step 1

Step 2

Step 3

Step 4

Solutions and Samples
of student work

5. **a.** Explanations will vary. Sample response:

   Chopping off a piece of the tetrahedron creates one more face, two more vertices, and three new edges; $1 + 2 - 3 = 0$. So, the net effect is no change.

   b. The result is always two. Students may use the same reasoning as for part **5a.** Sample response (using a table):

	Tetra-hedron	Step 1	Step 2	Step 3	Step 4
F	4	5	6	7	8
V	4	6	8	10	12
E	6	9	12	15	18
$F + V - E$	2	2	2	2	2

Hints and Comments

Materials models of a tetrahedron, optional (one per student)

Overview Students investigate the steps for cutting a diamond shaped like a tetrahedron and apply Euler's formula for each of the steps.

Planning Problem **5** may be done individually. You may wish to assign this problem as homework. Be sure to discuss students' answers.

Comments about the Problems

5. **Homework** This problem may be assigned as homework.

   a. As in problem **4,** students may discover that every time a tip is cut off, there are one new face, two new vertices, and three new edges.

   b. Encourage students to organize their information in a table.

Summary

There is a relationship between the numbers of faces, vertices, and edges of any polyhedron. This relationship is expressed in Euler's formula, $F + V - E = 2$. If you cut the corners from a polyhedron, you can investigate Euler's formula and see how the numbers of faces, vertices, and edges change.

Summary Questions

6. The net for a package is shown below.

    a. Draw the shape of the assembled package.

    b. Is the formula $F + V - E = 2$ true for the package?

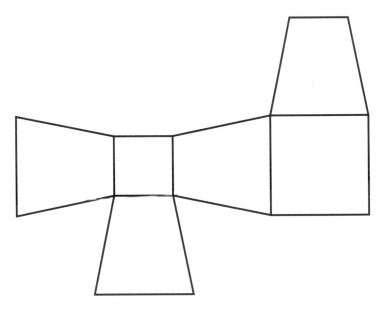

7. Picture a pyramid that has between 100 and 200 edges.

    a. Choose the number of edges for your pyramid (between 100 and 200). Is Euler's formula still valid for the pyramid you chose? Explain.

    b. Is Euler's formula valid for a pyramid with any number of edges between 100 and 200? Explain.

6. a.

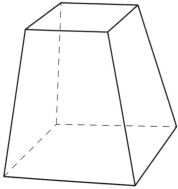

b. Yes, the formula holds.
$F = 6$, $V = 8$, $E = 12$
$6 + 8 - 12 = 2$

7. a. The formula will be valid for any number of edges that a pyramid can have. No pyramid can have an odd number of edges. Sample student response:

If the number of edges is 150, there are 75 edges in the base and 75 edges from the base to the top. $E = 150$; $F = 75 + 1 = 76$ (75 exposed faces plus the base); $V = 75 + 1$ (75 vertices at the base, plus one vertex at the top). Apply Euler's formula: $76 + 76 - 150 = 2$. It works!

b. No. It only applies for even numbers, because a pyramid cannot have an odd number of edges.

Materials straightedges or rulers (one per student); scissors, optional (one pair per student)

Overview Students read the Summary, which reviews the main concepts covered in this section. Then they draw the shape of the polyhedron that could be made from a given net and apply Euler's formula. They also apply Euler's formula to a pyramid with a large number of edges.

About the Mathematics Euler's formula is valid for all convex polyhedra.

Planning Students may work individually on problems **6** and **7**. These problems may also be used as informal assessments. After students complete Section F, you may assign appropriate activities from the Try This! section, located on pages 40–43 of the *Packages and Polygons* Student Book, for homework.

Comments about the Problems

6. Informal Assessment This problem assesses students' ability to understand and use Euler's formula, to draw two- and three-dimensional figures, and to develop spatial visualization skills.

Encourage students to do without a three-dimensional model. However, if students have difficulty visualizing, you might have them trace the net, cut it out, and assemble it before drawing it.

7. Informal Assessment This problem assesses students' ability to understand and use Euler's formula; to develop efficient counting strategies, involving geometric solids, that can be generalized; and to solve problems involving geometric solids.
If students have difficulty with this problem, you might have them draw a pyramid with a small number of edges to use as a reference.

Students should discover that the number of edges in a pyramid is always even. A pyramid has a base shaped like a polygon. For every side of that polygon, there is an edge (reaching from the base to the top of the pyramid). So, the total number of edges of a pyramid is twice the number of edges at the base.

Work Students Do

In this section, students construct five semi-regular polyhedra. They explore semi-regular polyhedra by considering the example of a soccer ball. Students apply Euler's formula to the five semi-regular polyhedra as well. They build a triangular box and apply Euler's formula to it. In doing so, students revisit side views, front views, and top views of different-shaped boxes. Students end the unit by designing their own packages in the shapes of polyhedra.

Goals

Students will:

- recognize and identify geometric shapes and structures in real objects and in representations;
- understand and use Euler's formula;
- recognize and identify properties of regular polygons and polyhedra;
- construct geometric models;
- draw two- and three-dimensional figures;
- make connections between different views of geometric solids;
- develop efficient counting strategies, involving geometric solids, that can be generalized;*
- develop spatial visualization skills.

  * This goal is assessed in other sections of the unit.

Pacing

- approximately two or three 45-minute class sessions

Vocabulary

- semi-regular polyhedron

About the Mathematics

The relationship between regular and some semi-regular polyhedra is the focus of this section. A semi-regular polyhedron is a solid made by cutting corners from a regular polyhedron so that new faces, all regular polygons, are created. In a regular polyhedron, all of the faces are congruent polygons; in a semi-regular polyhedron, the faces are two different sets of congruent polygons. When students create semi-regular polyhedra from regular polyhedra, they may begin to see a pattern. For instance, when a tetrahedron is cut from one corner of a cube, two vertices are added, three edges are added, and one face is added to the cube (see page 82 of the Teacher Guide). When a tetrahedron is cut from each of the eight corners, a semi-regular polyhedron is formed with triangular and octagonal faces. The number of faces is now 14, the number of vertices is 24, and the number of edges is 36, so Euler's formula still holds for this semi-regular polyhedron.

This section also ties together the concepts from other sections: nets, top and side views, building three-dimensional shapes, applying Euler's formula, and reasoning in three dimensions.

Materials

- Student Activity Sheets 10–13, pages 120–123 of the Teacher Guide (one of each per student)
- posterboard, pages 83, 89, 91, and 95 of the Teacher Guide (one sheet per student)
- straightedges or rulers, pages 83, 89, 91, and 95 of the Teacher Guide (one per student)
- scissors, pages 83, 89, 91, and 95 of the Teacher Guide (one pair per student)
- soccer ball, pages 85 and 87 of the Teacher Guide, optional (one per class)
- rubber bands, page 89 of the Teacher Guide, optional (240 per group of students)
- hand-held hole punch, page 89 of the Teacher Guide, optional (one per group of students)
- glue, pages 89, 91, and 95 of the Teacher Guide (one dispenser per group of students)
- tape, page 89 of the Teacher Guide, optional (one dispenser per group of students)
- colored pencils or crayons, pages 89 and 95 of the Teacher Guide, optional (one box per group of students)
- triangular-shaped boxes made by students on page 37 of the Student Book, page 93 of the Teacher Guide (two per group of students)

Planning Instruction

To introduce this section, you may want to have students count the numbers of faces, vertices, and edges on a soccer ball. Then ask students, *Does Euler's formula work for a soccer ball?* [yes] You may also wish to discuss polyhedra other than the five semi-regular polyhedra given on page 34 of the Student Book. For example, a pyramid with a hexagonal base has seven faces, seven vertices, and 12 edges, so Euler's formula holds. A triangular right prism has five faces, six vertices, and nine edges—again, Euler's formula holds. You may also want to have students refer to page 2 of the Student Book for other examples.

Students may work on problems 1, 2, and 7 individually. Problems 3, 5, and 11–13 may be done individually or in small groups. Students may work on the remaining problems in small groups.

The activity on page 36 of the Student Book, page 88 of the Teacher Guide, is optional. If time is a concern, you may omit this activity or assign it as homework.

Homework

Problems 3 (page 84 of the Teacher Guide), 7 (page 90 of the Teacher Guide), and 11–13 (page 94 of the Teacher Guide) may be assigned as homework. After students complete Section G, you may assign appropriate activities from the Try This! section, located on pages 40–43 of the *Packages and Polygons* Student Book. The Try This! activities reinforce the key mathematical concepts introduced in this section.

Planning Assessment

- Problem 11 may be used to informally assess students' ability to recognize and identify geometric shapes and structures in real objects and in representations, construct geometric models, draw two- and three-dimensional figures, make connections between different views of geometric solids, and develop spatial visualization skills.
- Problem 12 may be used to informally assess students' ability to recognize and identify properties of regular polygons and polyhedra.
- Problem 13 may be used to informally assess students' ability to understand and use Euler's formula.

G. SEMI-REGULAR POLYHEDRA

Activity

HOW TO SLICE A CUBE

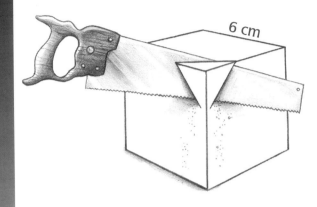

6 cm

4.25 cm   **?**   4.25 cm

4.25 cm

On **Student Activity Sheet 10** there is a net for a cube with six-centimeter edges, but part of the cube has been cut off.

1. Trace the net onto heavy paper and fold it into a cube.

   a. Describe your cube.

   b. Look through the hole in your cube and draw what you see.

2. Unfold your cube and cut the other corners off in the same manner. This results in a polyhedron with two different kinds of faces.

   a. What are the names of these faces?

   b. Draw a net for your polyhedron.

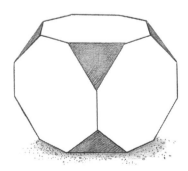

1. a. Answers will vary. Sample student response:

It looks like a cube with a hole in it because one of the corners is cut off. Nine edges are still 6 centimeters. When the corner was cut off, three edges became 4.25 centimeters and three new edges were created, which are about 2.5 centimeters each. The cube now has seven faces instead of six; it has 10 vertices instead of eight; and it has 15 edges instead of 12.

b. Drawings will vary, depending on how students look through the hole. Sample drawing:

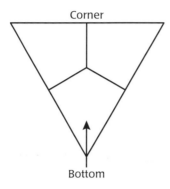

2. a. a regular triangle and regular hexagon

b. Nets may vary. Sample net:

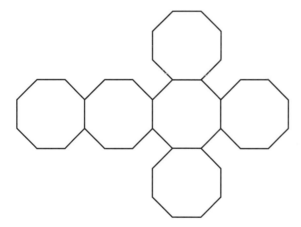

Materials Student Activity Sheet 10 (one per student); posterboard, (one sheet per student); straightedges or rulers (one per student); scissors (one pair per student)

Overview Students look through a hole in a cube and draw what they see. They cut off other corners in the same manner and design a net for the new polyhedron.

About the Mathematics The dimensions for the new edges that are created when a corner is cut off of a cube can be found in the following manner (but do not ask this from your students):

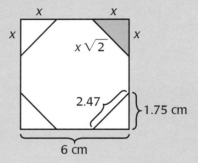

If you cut a length x off the edge of the cube, the length of the edge of the regular hexagon is $x\sqrt{2}$. $1.75\sqrt{2} = 2.47$

Planning Students may work on problems **1** and **2** individually. Before students begin problem **1,** you may want to tell them that they will have to unfold the cube for problem **2.** Discuss students' answers.

Comments about the Problems

1. Ask students to describe the modified cube as extensively as possible. Ask, *How many edges does the shape have? How long are the edges? How many faces does the shape have?*

2. The new cube with the other corners cut off should look like the shape at the bottom of page 33 of the Student Book.

  a. Students should note that the shape of each face is regular.

  b. Students may refer to the net from Student Activity Sheet 10 as they draw the net of the new polyhedron.

Soccer Balls

A soccer ball is similar to a sphere with a diameter of about 22 centimeters. The ball in the photograph is made of black and white pieces of leather.

3. a. There are differences, besides color, between the black and the white faces. What other differences do you see?

b. On the soccer ball pictured above, you can see six black pieces. How many black pieces do you think are on the whole soccer ball?

c. How many white pieces are on the whole soccer ball?

3. a. The black faces are shaped like pentagons, and the white faces are shaped like hexagons. The black faces are smaller than the white ones.

b. There are 12 black pieces. Explanations will vary. Students may say that because they see six black pentagons, there must be six on the other side.

c. There are 20 white pieces. Explanations will vary. Sample student explanation:

> I see 10 white faces, so I guess there must be 20 white faces in total. I can check my answer as follows:
>
> There are 12 black pentagons. Around each pentagon there are five hexagons;
> $5 \times 12 = 60$. But now I have counted each hexagon three times because each of them is connected to three different pentagons. So, I should divide 60 by three. The answer is 20.

Materials soccer ball, optional (one per class)

Overview Students investigate the shapes of the faces of a soccer ball. They determine the number of each sort of shape.

About the Mathematics There is a nice way of proving that every polyhedron consisting of pentagons and hexagons must have 12 pentagons. But do not expect this proof from your students.

Suppose:
the number of pentagons = a
the number of hexagons = b
so $F = a + b$

For each pentagon, there are five vertices ($5a$) and for each hexagon there are six vertices ($6b$). The sum $5a + 6b$, however, counts each vertex three times (each vertex touches three polygons) so the result is divided by three.

$$V = \frac{(5a + 6b)}{3} = \frac{5a}{3} + 2b$$

For each pentagon, there are five edges ($5a$), and hexagons have six ($6b$). The sum counts each edge twice, so the result is divided by two.

$$E = \frac{(5a + 6b)}{2} = \frac{5a}{2} + 3b$$

Using Euler's formula:

$$F + V - E = 2$$
$$(a + b) + (\frac{5}{3a} + 2b) - (\frac{5}{2a} + 3b) = 2$$
$$a + \frac{5}{3a} - \frac{5}{2a} + b + 2b - 3b = 2$$
$$\frac{8}{3a} - \frac{5}{2a} = 2$$
$$\frac{1}{6a} = 2$$
$$a = 12$$

Planning It is helpful to have an actual soccer ball for students to see. Students may work on problem **3** individually or in small groups. This problem can also be assigned as homework. If students have difficulty, you may wish to discuss their answers in class.

Comments about the Problems

3. Homework This problem may be assigned as homework.

a. Encourage students to note as many differences as possible. If students have difficulty, you might have them count the edges of each shape or compare the shapes on the real soccer ball.

c. Encourage students to give an explanation for their answer.

Icosahedra and Soccer Balls

4. a. The shape on the right is an icosahedron. Explain how a soccer ball is related to an icosahedron.

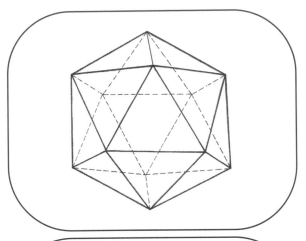

b. Tim gives the following reasoning:

A soccer ball has 12 black pentagons. Each pentagon borders five white hexagons. Therefore, the number of hexagons has to be 12 × 5 = 60.

What is Tim's mistake?

c. Investigate to see whether Euler's formula is valid for the soccer ball.

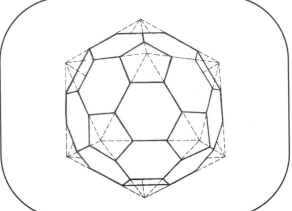

4. a. Answers will vary. Sample student response:

A soccer ball is made by taking an icosahedron and cutting off the same amount at each vertex. This leaves a pentagon where there used to be a vertex, forming five additional edges. There are 12 vertices in an icosahedron, so there will be 12 pentagons and 60 additional edges.

b. Each white hexagon touches three black pentagons, so Tim counted each hexagon three times; $60 \div 3 = 20$ hexagons.

c. Euler's formula is valid for the soccer ball. Explanations will vary. Sample student explanation:

$F = 20 + 12 = 32$
$V = 60$
$12 \times 5 = 60$
$20 \times 6 = 120$
$60 + 120 = 180$
But each vertex belongs to three faces, so I counted them three times; $180 \div 3 = 60$. So there are 60 vertices.

$E = 90$
$12 \times 5 = 60$
$20 \times 6 = 120$
$60 + 120 = 180$
But this way, I counted all the edges twice; $180 \div 2 = 90$. So there are 90 edges.

$F + V - E = 32 + 60 - 90 = 2$

Materials soccer ball, optional (one per class)

Overview Students further investigate a soccer ball. They relate it to an icosahedron, they reason about the number of hexagons on a soccer ball, and they check Euler's formula for the soccer ball.

Planning Students may work on problem **4** in small groups. Discuss their answers.

Comments about the Problems

4. a. If students have difficulty seeing and describing the relationship, you might suggest that the soccer ball can be made by cutting off parts from the icosahedron. Encourage them to put into words what this will do to the shape. Students may find it helpful to draw several stages of the process of cutting off the corners of the icosahedron.

b. Some students may already have built on this reasoning while answering problem **3,** on page 34 of the Student Book. (See Solutions and Samples, problem **3c.**)

c. If students do not remember Euler's formula, they may look it up again. They will have to determine the number of faces, edges, and vertices before they can apply the formula. If students have difficulty, you might remind them that they must avoid counting anything twice. Bordering edges and shared vertices should be counted only once.

Interdisciplinary Connection Did you know that the 1996 Nobel Prize in Chemistry went to three chemists who discovered a new form of molecular carbon that is shaped like a soccer ball? In modern chemistry, the C_{60} molecule, shown below, is called a "buckminster fullerene," or "buckyball." Buckyballs were named after Buckminster Fuller, who developed the geodesic dome, a very strong structure used in architecture. A geodesic dome also is shaped like a soccer ball.

C_{60}

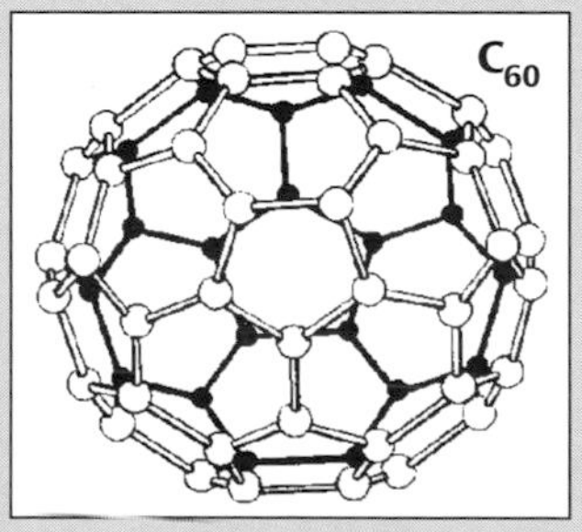

Five Semi-regular Polyhedra

Activity

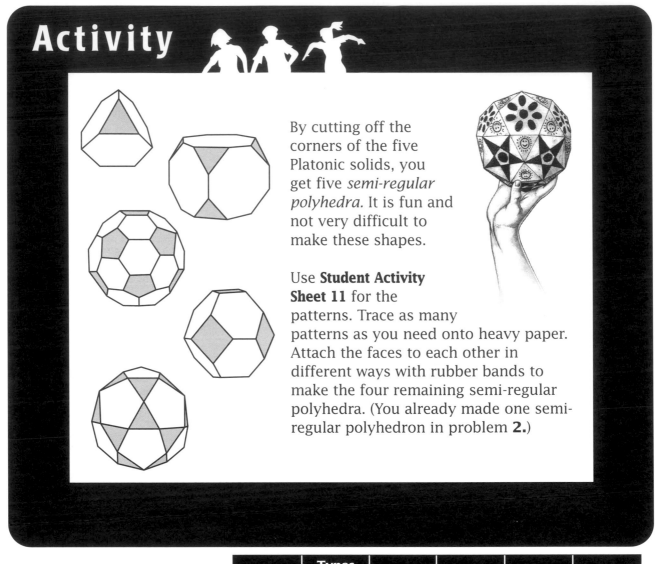

By cutting off the corners of the five Platonic solids, you get five *semi-regular polyhedra.* It is fun and not very difficult to make these shapes.

Use **Student Activity Sheet 11** for the patterns. Trace as many patterns as you need onto heavy paper. Attach the faces to each other in different ways with rubber bands to make the four remaining semi-regular polyhedra. (You already made one semi-regular polyhedron in problem **2.**)

5. Study your semi-regular polyhedra carefully to find *F*, *V*, and *E*. Check your results using Euler's formula and fill in the table on **Student Activity Sheet 12.**

Shape	Types of Faces	F	V	E	F+V−E
	Triangle and Hexagon			18	
		14			
			60		

5.

Shape	Types of Faces	F		V	E	F + V − E
	Triangle and Hexagon	4 4 } 8		$\frac{12 + 24}{3} = 12$	$\frac{12 + 24}{2} = 18$	8 + 12 − 18 = 2
	Triangle and Octagon	8 6 } 14		$\frac{24 + 48}{3} = 24$	$\frac{24 + 48}{2} = 36$	14 + 24 − 36 = 2
	Square and Hexagon	6 8 } 14		$\frac{24 + 48}{3} = 24$	$\frac{24 + 48}{2} = 36$	14 + 24 − 36 = 2
	Pentagon and Hexagon	12 20 } 32		$\frac{60 + 120}{3} = 60$	$\frac{60 + 120}{2} = 90$	32 + 60 − 90 = 2
	Triangle and Pentagon	20 12 } 32		$\frac{60 + 60}{4} = 30$	$\frac{60 + 60}{2} = 60$	32 + 30 − 60 = 2

Note: The table on Student Book page 36 has an incorrect entry in the last row, under vertices. The number 60 belongs in the Vertices column for the soccer ball. This has been corrected on Student Activity Sheet 12.

Materials Student Activity Sheets 11 and 12 (one of each per student); posterboard, optional (one sheet per student); straightedges or rulers (one per student); scissors (one pair per student); rubber bands, optional (240 per group); hand-held hole punch, optional (one per group); glue, optional (one dispenser per group); tape, optional (one dispenser per group); colored pencils or crayons, optional (one box per group)

Overview Students study five semi-regular polyhedra and determine the numbers of faces, edges, and vertices for each. They check their results using Euler's formula.

Planning Students may work on the activity and problem **5** individually or in small groups. The activity is optional. If time is a concern, you may omit the activity or assign it as homework.

Comments about the Problems

5. Students may be able to fill in the chart without having the actual models available to look at by using the following strategy:

The first shape has four triangular faces and four hexagonal faces, or eight faces all together. There are 4 × 3 vertices from the four triangles and 4 × 6 vertices from the four hexagons. However, in order to eliminate duplication, you must divide the total (12 + 24) by three, since each vertex is shared by three faces.

There are 4 × 3 edges for the triangles and 4 × 6 edges for the hexagons; in order to eliminate duplication, the total must be divided by two, because each edge is shared by two faces.

Getting It All Together

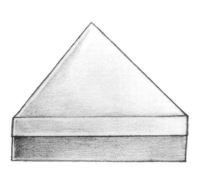

If you look again at the packages on page 1, you will notice a triangular box and a hexagonal box.

6. How are the shapes of the two boxes related?

On the right is a net for the triangular box. Two larger copies of this net appear on **Student Activity Sheet 13.**

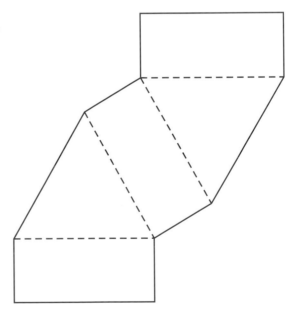

7. a. On **Student Activity Sheet 13,** draw the glue strips that are necessary to build the boxes.

   b. Trace your patterns onto heavy paper and build the boxes.

8. How can you make the hexagonal prism shown above with the triangular prisms that you just made? Explain your answer with a drawing.

6. Answers will vary. Sample responses:

The two shapes have the same height.

They both have the shape of a prism.

7. a. There are several possibilities. One way to draw the glue strips is shown below:

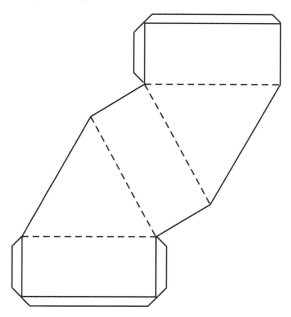

b. Students' prisms should look like the one on the upper left corner of page 37 of the Student Book.

8. The hexagonal prism can be made by putting six of the triangular prisms together, as shown in the top-view drawing below:

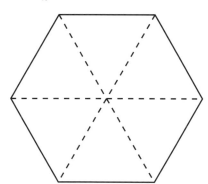

Materials Student Activity Sheet 13, (one per student); posterboard (one sheet per student); straightedges or rulers (one per student); scissors (one pair per student); glue (one dispenser per student)

Overview Students describe the relationship between a triangular box and a hexagonal box. They draw glue strips on the nets and build the boxes.

Planning Students may work on problem **6** in small groups, but you may wish to discuss the problem in class. Problem **7** can be done individually in class or assigned as homework. Problem **8** may be done in small groups and discussed in class.

Comments about the Problems

6. Students will not be able to tell that the top of the triangular box is regular just by looking at the drawing on page 37 of the Student Book. Instead, they should refer to the net on Student Activity Sheet 13.

7. Homework This problem may be assigned as homework. Students can draw the glue strips on Student Activity Sheet 13. You may have students reason about where to put glue strips before they cut out the nets and build the boxes. If students have difficulty, you might remind them that the box has a top and a bottom. They can check the position of the glue strips by building the box. Some students may need extra copies of Student Activity Sheet 13 because of cutting or gluing errors.

8. Some students may need to put the six triangular boxes together to make the hexagonal box before they are able to make a drawing. A top-view drawing is the easiest way to show how the triangular boxes should be put together, but students may make any drawing they wish.

9. Put two triangular boxes together in order to form the box shown below.

 a. Precisely describe the box you have made.

 b. Draw a net of this box.

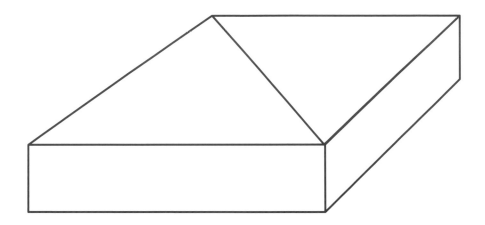

10. a. Copy and complete the table on the right to find out how the numbers of edges, vertices, and faces change as you add triangular boxes to a shape.

b. Describe any patterns you find in the table.

Box	Faces	Vertices	Edges	$F+V-E$
△	5	6	9	2
▱	6	8		
⬠				
⬡				
⬡				
⬡				

9. a. Answers will vary. Sample responses:

The box is symmetrical.
It looks like a diamond.
In a top view, all sides have the same length.
The top view looks like a rhombus because it has four equal sides.
The shape of the top view has two angles measuring less than 90 degrees and two angles measuring more than 90 degrees.
The large angles in the top view are twice the size of the small angles.

b. Nets will vary. Sample net:

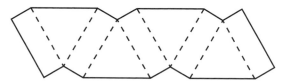

10. a.

Box	Faces	Vertices	Edges	F + V − E
	5	6	9	2
	6	8	12	2
	6	8	12	2
	8	12	18	2
	9	14	21	2
	8	12	18	2

b. There are no obvious patterns in the table. For several boxes, the number of faces increases by one; the number of vertices increases by two; the number of edges increases by three. However, the third and the sixth box in the list do not fit this pattern. Euler's formula holds for all the boxes.

Materials triangular boxes made by students on page 37 of the Student Book (two per group of students)

Overview Students make a hexagonal box with the triangular boxes they built in problem **7.** They draw a net for a box made of two of the triangular boxes. They also fill in a table with the number of faces, edges, and vertices for a box made of two, three, four, five, and six of the triangular boxes and check Euler's formula in each case.

Planning Students may work on problems **9** and **10** in small groups. Discuss their answers in class.

Comments about the Problems

9. a. Encourage students to describe as many characteristics of the solid as possible.

b. Students should compare their drawings.

10. a. Students may fill in the table differently. For example, for the box in the second row, its top and bottom can be said to have two faces each, to make a total of eight faces. The entries in the second row would be: Faces: 8; Vertices: 8; and Edges: 14. Notice that Euler's formula still works (8 + 8 = 16; 16 − 14 = 2).

b. You may have to remind students that Euler's formula states that $F + V - E = 2$. Ask students, *Does Euler's formula hold for each case?* [Yes.]

Ask students, *How can you explain why the boxes in the list do not fit a pattern?* [Adding a new box to the structure does not change the number of tops and bottoms. Usually only the number of sides increases, but adding the third box does not add an extra side. Adding the sixth box decreases the number of sides.]

Summary

Semi-regular polyhedra have at least two different regular polygons as faces. These solids can be made by cutting pieces off regular polyhedra. For example, if you cut pieces off the corners of an icosahedron, you get a semi-regular polyhedron made of pentagons and hexagons.

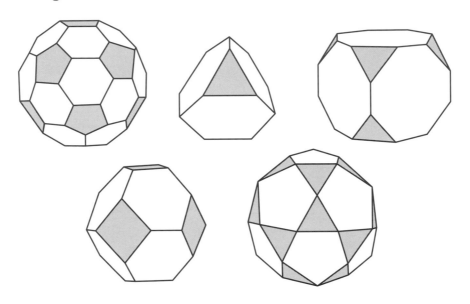

Summary Questions

11. Suppose that you have invented a new toy.

    a. Design a unique package for your toy in the shape of a polyhedron.

    b. Make a net for your package. Build and decorate your package.

    c. Draw a top view and a precise front view of your package.

12. Is your package a regular polyhedron, a semi-regular polyhedron, or neither? Explain.

13. Show that Euler's formula works for your package.

11. a. Designs will vary. A possible design with measurements is shown here:

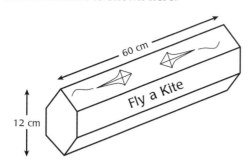

It is a box designed to hold a kite. The kite will be rolled up to be put in the box; the box is as long as the longest stick of the kite.

b. Nets will vary. The net of the sample design for problem **11a** can appear as follows:

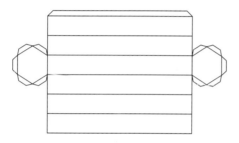

Decorations will vary.

c. Answers will vary. The top view and front view of the sample design for problem **11a** look like the following:

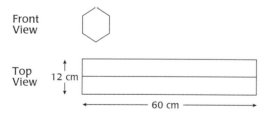

12. Answers will vary. The sample design for problem **11** is neither a regular nor a semi-regular polyhedron. It is not regular, because there are two different shapes for the faces. It is not semi-regular, because it cannot be made by cutting off pieces from a regular polygon. The top and the bottom are hexagons (regular polygons) of the same size; the side faces are rectangles (regular polygons) of the same size.

13. Answers will vary. The formula works for all polyhedra. In case of the sample design in problem **11,** there are eight faces, 12 vertices, and 18 edges. So, $F + V - E = 8 + 12 - 18 = 2$. The formula works.

Materials posterboard (one sheet per student); straightedges or rulers (one per student); scissors (one pair per student); glue (one dispenser per group); colored pencils or crayons, optional (one box per group)

Overview Students read and discuss the Summary. They design a unique package in the shape of a polyhedron, make a net for it, build the package, and draw top and front views. Students describe their packages in terms of their characteristics as polyhedra. Then, they show that Euler's formula holds for their designs.

Planning Problems **11–13** can be done individually or in small groups. They can also be assigned as homework. Together they can be used as an informal assessment. After students complete Section G, you may assign appropriate activities from the Try This! section, located on pages 40–43 of the *Packages and Polygons* Student Book, for homework.

Comments about the Problems

11–13. Homework These problems may be assigned as homework.

11. Informal Assessment This problem assesses students' ability to recognize and identify geometric shapes and structures in real objects and in representations. It also assesses students' ability to construct geometric models, draw two- and three-dimensional figures, make connections between different views of geometric solids, and develop spatial visualization skills.

a. Students should indicate the measurements of their designs.

b. Students should pay attention to the placement of glue strips.

c. Some students may need to construct the actual package and look at it from the top and the front in order to be able to make these drawings.

12. Informal Assessment This problem assesses students' ability to recognize and identify properties of regular polygons and polyhedra.

13. Informal Assessment This problem assesses students' ability to understand and use Euler's formula.

Assessment Overview

Students work on five assessment activities that you can use to collect additional information about what each student knows about polygons, polyhedra, and Euler's formula.

Goals

- recognize and identify geometric shapes and structures in real objects and in representations

- understand and use Euler's formula

- use the relationships between angles and turns to solve problems

- recognize and identify properties of regular polygons and polyhedra

- construct geometric models

- draw two- and three-dimensional figures

- develop efficient counting strategies, involving geometric solids, that can be generalized

- reason about the structure of the Platonic solids

- develop spatial visualization skills

- solve problems involving geometric solids

Assessment Opportunities

Naming Shapes, Polyhedra Constructions, Alternative Assessment

Tiling with Polygons, Alternative Assessment

Tiling with Polygons

Naming Shapes, Polyhedra Constructions, Tiling with Polygons, Alternative Assessment

Naming Shapes, Alternative Assessment

Naming Shapes, Alternative Assessment

Tiling with Polygons, Hidden Faces

Hidden Faces, Alternative Assessment

Naming Shapes, Polyhedra Constructions, Hidden Faces, Alternative Assessment

Hidden Faces, Alternative Assessment

Pacing

When combined, the five assessment activities will take approximately two 45-minute class sessions. See the Planning Assessment section for further suggestions as to how you might use the assessment activities.

About the Mathematics

The five assessment activities evaluate the major goals of the *Packages and Polygons* unit. Refer to the Goals and Assessment Opportunities section on the previous page for information regarding the specific goals assessed in each assessment activity.

Materials

- Assessments, pages 124–128 of the Teacher Guide (one of each per student)
- grid paper, page 99 of the Teacher Guide, optional (one sheet per student)
- scissors, pages 99 and 107 of the Teacher Guide (one pair per student)
- glue or tape, pages 99 and 107 of the Teacher Guide (one dispenser per student)
- red and blue pencils, pages 101 and 107 of the Teacher Guide (two per student)
- paper models of cubes, pages 103 and 105 of the Teacher Guide, optional (one per student)
- models of polyhedra made by students on page 24 of the Student Book, page 107 of the Teacher Guide (one set per group of students)
- tracing paper, page 107 of the Teacher Guide (several sheets per student)
- rulers, page 107 of the Teacher Guide (one per student)
- posterboard, page 107 of the Teacher Guide (several sheets per group of students)
- straws and string or pipe cleaners, page 107 of the Teacher Guide (50 segments per group of students)

Planning Assessment

You may want students to work on these assessments individually so that you can evaluate each student's understanding and abilities. Make sure that you allow enough time for students to complete the assessment activities. Students are free to solve each problem in their own way. They may choose to use any of the models introduced and developed in this unit to solve problems that do not call for a specific model.

Scoring

Answers are provided for all assessment problems. The method of scoring the problems depends on the types of questions in each assessment. Most questions require students to explain their reasoning or justify their answers. For these questions, the reasoning used by the students in solving the problems as well as the correctness of the answers should be considered as part of your grading scheme. A holistic scoring scheme can be used to evaluate an entire task. For example, after reviewing a student's work, you may assign a key word such as *emerging, developing, accomplishing,* or *exceeding* to describe his or her mathematical problem-solving, reasoning, and communication.

On other tasks, it may be more appropriate to assign point values for each response. Students' progress toward the goals of the unit should also be considered. Descriptive statements that include details of a student's solution to an assessment activity can be recorded. These statements would provide insight into a student's progress toward a specific goal of the unit. Descriptive statements can be more informative than recording only a score and can be used to document students' growth in mathematics over time.

NAMING SHAPES

Use additional paper as needed.

1. Name the three-dimensional shapes you recognize in the objects pictured below.

    a.

    b.

    c.
    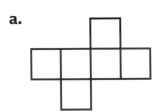

    d.

2. How would you explain to someone how to tell the difference between a pyramid and a prism?

3. Is it possible to make a cube from any of the nets pictured below?

    a.    b. 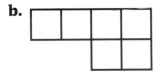   c.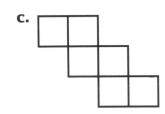

4. The cube below is painted half gray and half white. Draw two different nets for this cube. Be sure to color the faces.

1. a. spheres (marbles)

  b. pyramid (top of building)
  rectangular prism (lower part)

  c. cone (ice cream cone)

  d. prism (garden shed)

2. Answers will vary. Sample student response:

Put a prism and a pyramid on a table. Look at the edges that go up. The edges of the prism go straight up. They do not get closer to each other. The edges of the pyramid come together in one point as they go up.

3. a. yes

  b. no

  c. yes

4. Nets will vary. Sample nets:

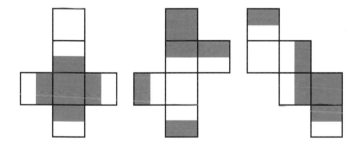

Materials Naming Shapes assessment, page 124 (one per student); grid paper, optional (one sheet per student); scissors, optional (one pair per student); glue or tape, optional (one dispenser per group)

Overview Students name the shapes they recognize in the pictures. They decide whether or not a cube can be made from the given nets. Then they draw two nets for a two-colored cube.

About the Mathematics These assessment activities evaluate students' ability to recognize and identify geometric shapes and structures in real objects and in representations. They also evaluate students' ability to recognize and identify properties of regular polygons and polyhedra, construct geometric models, draw two- and three-dimensional figures, and develop spatial visualization skills.

Planning You may choose to do this assessment any time after completing Section B. You may want students to work on these assessment problems individually. Problem **2** may be assigned as homework.

Comments about the Problems

  1. b. Some students may mention only the shape of the top of the structure. Other students may realize that this structure can be seen as a composite of two shapes.

  d. Some students may have difficulty recognizing a prism. Some students may also see this structure as a composite of two structures: a right rectangular prism (bottom) and a prism (roof).

  2. Homework This problem can be assigned as homework. You may want to ask students to explain this problem to their families and write a short report in their notebooks.

  3–4. Allow students to draw the nets on a separate sheet of paper and cut them out to verify their answers.

POLYHEDRA CONSTRUCTIONS

Use additional paper as needed.

1. **a.** The structure on the right can be made of wood. What shapes do you have to glue together to make this structure? How many do you need?

   b. With what shape could you fill the corners to make the structure into a regular polyhedron?

   c. If you fill all the corners, what shape will the whole structure have?

2. The drawings on the right are of models of polyhedra.

   a. Figure i is built from two tetrahedra. Color one tetrahedron red and the other one blue.

i.

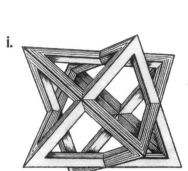

   b. From which two polyhedra is figure ii constructed? Color each one in a different color.

ii.

   c. Figure iii is constructed from three regular polyhedra. Color one of these. What is the name of this polyhedron?

iii.

1. a. six right rectangular prisms with sections missing from the center

b. with a cube

c. a cube

2. a.

b. two cubes

c. Figure iii is constructed from three octahedra. Students may have colored any one of the following polyhedra:

or   or

Materials Polyhedra Constructions assessment, page 125 (one per student); red and blue pencils (two per student)

Overview Students recognize shapes in a wooden structure. Then they recognize regular polyhedra in more complex models.

About the Mathematics These assessment problems evaluate students' ability to recognize and identify geometric shapes and structures in real objects and in representations. They also evaluate students' ability to recognize and identify properties of regular polygons and polyhedra to develop spatial visualization skills.

Planning You may want students to work on these assessment problems individually.

Comments about the Problems

1. This problem asks students to visualize this shape in order to find the number of rectangular prisms that can be used to build the shape. Note that the inside of the shape might be empty.

2. Students should be able to color the two tetrahedra in problem **2a.** Problems **2b** and **2c** are more difficult. Students' solutions may give you a good impression of their ability to recognize the shapes in a three-dimensional drawing.

TILING WITH POLYGONS

Use additional paper as needed.

1. This is a tiling composed of regular polygons.

   a. Name the regular polygons you recognize.

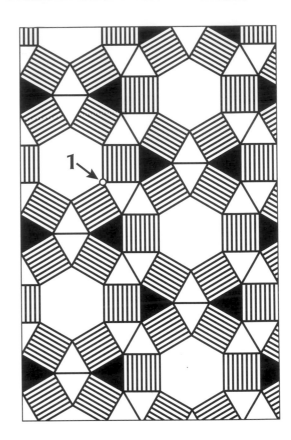

   b. Look at the point labeled 1. You can find four angles with a vertex at 1. What is the measure in degrees of each of these angles? Explain how you found your answer.

2. One tip is cut off of a cube. On the opposite vertex of the cube, the one that is hidden in the picture on the right, a similar piece will be cut off.

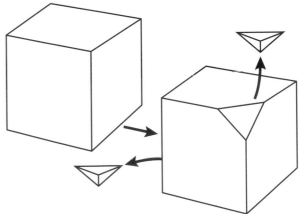

   a. How many vertices will the new shape have?

   b. How many faces?

   c. How many edges?

   d. In the space on the right, write a formula for the relationship between the numbers of faces, vertices, and edges using *F*, *V*, and *E*. Use your formula to check your answers to parts **a, b,** and **c.**

Solutions and Samples
of student work

1. **a.** hexagons, regular triangles, squares, 12-gons

   b. 60°, 90°, 90°, 120°

   Explanations will vary. Sample explanation:

   Squares have interior angles of 90°. Regular (equilateral) triangles have angles of 60°, and regular hexagons have angles of 120°.

   Students may also reason that the sum of all four angles is 360°, so the triangle and hexagon make 180° (subtracting the two right angles). The triangle has a 60° angle, so the hexagon measures 120°.

2. **a.** 12 vertices

   b. 8 faces

   c. 18 edges

   d. $F + V - E = 2$,
   $8 + 12 - 18 = 2$

Hints and Comments

Materials Tiling with Polygons assessment, page 126 (one per student); paper models of cubes, optional (one per student)

Overview Students name the regular polygons they see in a tessellation and find the measurements of the angles. Then they solve problems about a cube from which two tips are cut off and apply Euler's formula.

Planning You may want students to work on these assessment problems individually.

Comments about the Problems

1. Some students may find it confusing to return to a two-dimensional situation, a pattern of tiles.

   b. You may want to remind students to show how they arrived at their answers. They may use a variety of strategies to solve this problem.

2. If students have difficulty solving this problem, you may allow them to use a paper model of a cube.

HIDDEN FACES

Use additional paper as needed.

Pictured here is a view of a solid. The hidden half is identical to the visible half.

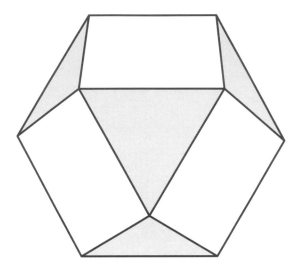

1. How can you make this shape from a cube?

2. Suppose you were to make a bar model of this shape. How many edges do you need? How many vertices do you need? Explain how you found your answers.

1. Cut the corners off the cube from the midpoint of each edge. The shaded parts were made from the faces of the original cube.

2. 24 edges; 12 vertices

   Strategies will vary. Some students may draw the model as shown below.

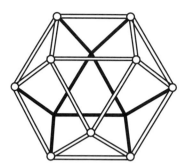

   Students may reason that there are six squares, each with 4 edges. This accounts for 24 edges. There are also 8 triangular faces, but each of these share all its edges with a square, so there are 24 edges. Students might count the number of vertices.

Materials Hidden Faces assessment, page 127 (one per student); paper models of cubes, optional (one per student)

Overview Students investigate a picture of a crystal and explain how it can be made from a cube. Then they explain how to make a bar model of the crystal.

Planning You may want students to work on these assessment problems individually. This assessment can also be assigned as homework or extra credit.

Comments about the Problems

1–2. Students' answers will show whether they understand how new shapes can be made by cutting off certain parts of existing shapes. If students find it difficult to visualize what parts to cut off the cube, you might allow them to use paper models.

ALTERNATIVE ASSESSMENT

Use additional paper as needed.

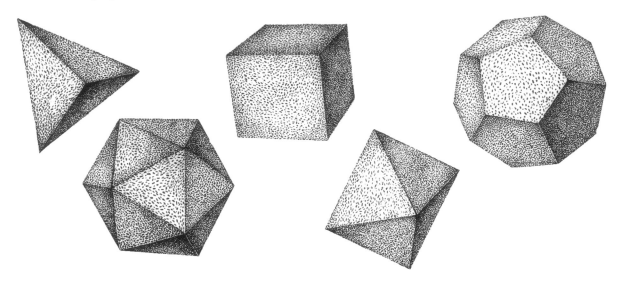

1. Your teacher has placed a number of solids at the front of the room. Your group will choose one or two of these solids to make a design. First, your group will decide what kind of object you are going to make. You may decide to design a package, an art object, an object that could be used at a playground, or something else. Next, choose one or two of these solids and make your design; you can paste solids together or cut off pieces so you will get semi-regular polyhedra.

2. Write a report about your design. Be sure that your report includes the following information:

   • the purpose of your design,

   • detailed information about the shape, and whether or not Euler's formula is true for it,

   • directions for making the design, including a net.

3. Present your design to the class.

1. Designs will vary. Sample design:

2. Reports will vary. Sample student response for the design in the solution to problem **1:**

This is a package for a pie. We started with a cube and cut it in half. Then we cut off the four upper corners as shown below.

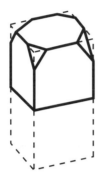

The bottom face has the shape of a square, four faces have the shape of a regular triangle, four faces have the shape of a hexagon, and one face has the shape of an 8-gon.

Applying Euler's formula:
$F = 6 + 4 = 10$
$V = 4 + 4 \times 3 = 4 + 12 = 16$
$E = 12 + 4 \times 3 = 24$
$F + V - E = 10 + 16 - 24 = 2$
So, the formula is true.

You can use the net below to make this package:

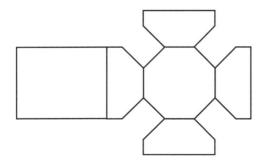

3. Presentations will vary.

Materials Alternative Assessment activity, page 128 (one per student); scissors (one pair per student); glue or tape (one dispenser per group); colored pencils (two per student); models of polyhedra made by students on page 24 of the Student Book (one set per group of students); tracing paper (several sheets per student); rulers (one per student); posterboard (several sheets per group); straws and string or pipe cleaners (50 segments per group)

Overview Students design an object using one or more of the Platonic solids. They investigate and describe their designs in detail. Finally, they make presentations of their designs.

About the Mathematics These assessment activities evaluate students' ability to recognize and identify geometric shapes and structures in real objects and in representations. They also evaluate students' ability to understand and use Euler's formula, recognize and identify properties of regular polygons and polyhedra, construct geometric models, and draw two- and three-dimensional figures. Finally, these activities assess students' ability to reason about the structure of the Platonic solids, develop spatial visualization skills, and solve problems involving geometric solids.

Planning You may want students to work on this assessment activity in small groups.

Comments about the Problems

1. Students must first reflect on the purpose of their design. Then they can use models of Platonic solids to make their design. Depending on the design, students may make a paper model or a bar model.

2. Students can make a drawing of the shape by tracing it from the unit or by enlarging a drawing in the unit. They can describe how the shape can be made out of a Platonic solid. Accept any answer that describes the shape geometrically.

Students can draw the shape and extend its edges with dotted line segments in a different color. This will create a picture of the Platonic solid from which the semi-regular polyhedra has been made as shown in the solutions column.

Students should include a net in their guidelines explaining how to make the design.

Glossary

The Glossary defines all vocabulary words listed on the Section Opener pages. It includes the mathematical terms that may be new to students, as well as words having to do with the contexts introduced in the unit. (*Note:* The Student Book has no glossary. This is in order to allow students to construct their own definitions, based on their personal experiences with the unit activities.)

The definitions below are specific to the use of the terms in this unit. The page numbers given are from this Teacher Guide.

edge (p. 30) a side of a polygon or a side of a face of a polyhedron

face (p. 22) a flat (two-dimensional) side of a three-dimensional solid

net (p. 16) a flat (two-dimensional) pattern to be cut and folded into a three-dimensional shape

Platonic solids (p. 62) the five regular polyhedra (tetrahedron, cube, octahedron, dodecahedron, and icosahedron)

polygon (p. 40) a closed two-dimensional figure made up of three or more linear sides

polyhedron (p. 58) a three-dimensional shape whose faces are polygons

regular polygon (p. 42) a polygon in which all angles are of the same measure (equiangular) and all sides are of the same length (equilateral)

regular polyhedron (p. 62) a polyhedron in which all faces are the same regular polygon and the same number of edges intersect at each of its vertices

semi-regular polyhedron (p. 88) a polyhedron that has two different regular polygons as faces

tetra- (p. 66) four

truncate (p. 8) to shorten by cutting off

vertex (p. 30) a point common to two sides of a polygon or to more than one edge of a polyhedron

Blackline Masters

Dear Family,

Your child will soon begin a study of two- and three-dimensional shapes in the *Mathematics in Context* unit *Packages and Polygons*. Below is a letter to your child that opens the unit, describing the unit and its goals.

The unit begins with an investigation into packages of various shapes. You might want to take your child to the grocery store and look at the variety of packages on the shelves. Discuss with your child why certain items are packaged in certain shapes. For example, milk cartons probably have triangular tops in order to help prevent spilling.

As your child works through the unit, point out different geometric shapes around your house. Look at the ways in which triangles are used to make some structures more stable. Have your child find the triangles in the legs of an ironing board or under a table, for example. Discuss with your child what would happen to the table if the triangles were not there.

You can also discuss the shapes of the buildings around your town or neighborhood. Have your child show you how to count the sides, edges, and corners of the shapes and explain how these numbers are related.

We hope you and your child enjoy learning about packages and polygons together!

Sincerely,

The Mathematics in Context Development Team

Dear Student,

Welcome to the unit *Packages and Polygons*.

Have you ever wondered why certain items come in differently shaped packages? The next time you are in a grocery store, look at how things are packaged. Why do you think table salt comes in a cylindrical package? Which packages do you think are the most practical?

Geometric shapes are everywhere. Look at the skyline of a big city. Can you see different shapes? Why do you think some buildings are built using one shape and some using another?

In this unit, you will explore a variety of two- and three-dimensional shapes and learn how they are related. You will build models of these shapes using heavy paper, or straws and pipe cleaners, or gumdrops and toothpicks. As you work through the unit, notice the shapes of objects around you. Think about how the ideas you are learning in class apply to those shapes.

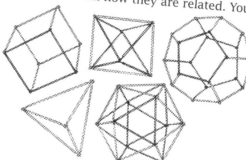

We hope you enjoy your investigations into packages and polygons.

Sincerely,

The Mathematics in Context Development Team

Cut out each of the following hexagonal lids.

A.

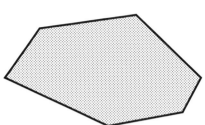

B.

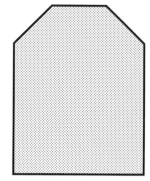

C.

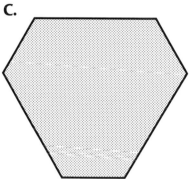

D.

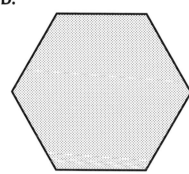

6. In how many ways does hexagon A fit on top of prism A on page 16? hexagon B on top of prism B? In how many ways do hexagons C and D fit on top of their prisms?

7. Compare the four hexagonal lids. How are their shapes similar? How do they differ?

Hexagon D is a special type of polygon; it is a *regular polygon.*

8. What do you think makes hexagon D "regular"?

Use with *Packages and Polygons,* page 19.

Name _____

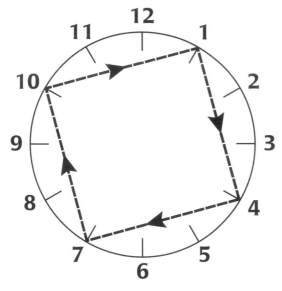

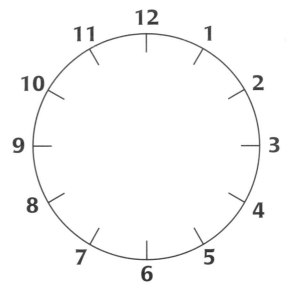

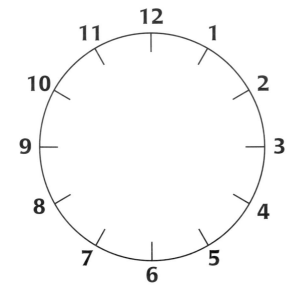

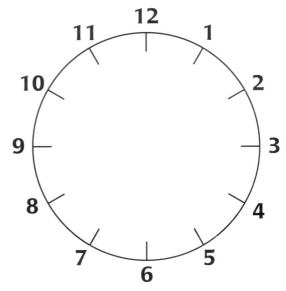

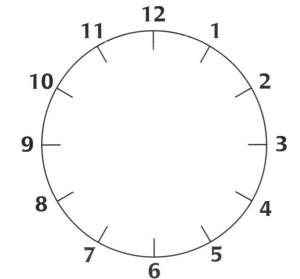

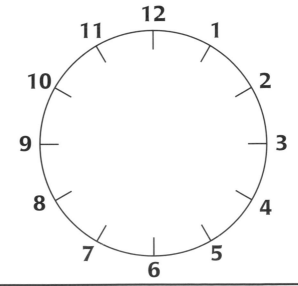

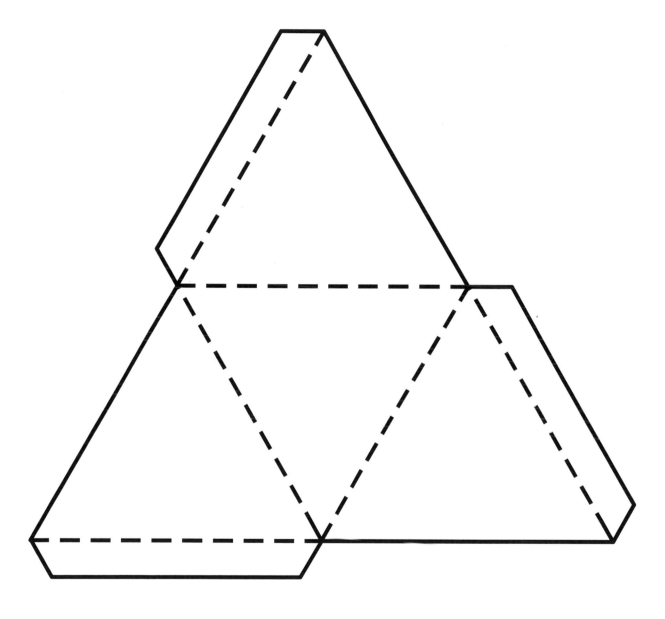

Use with *Packages and Polygons,* page 24.

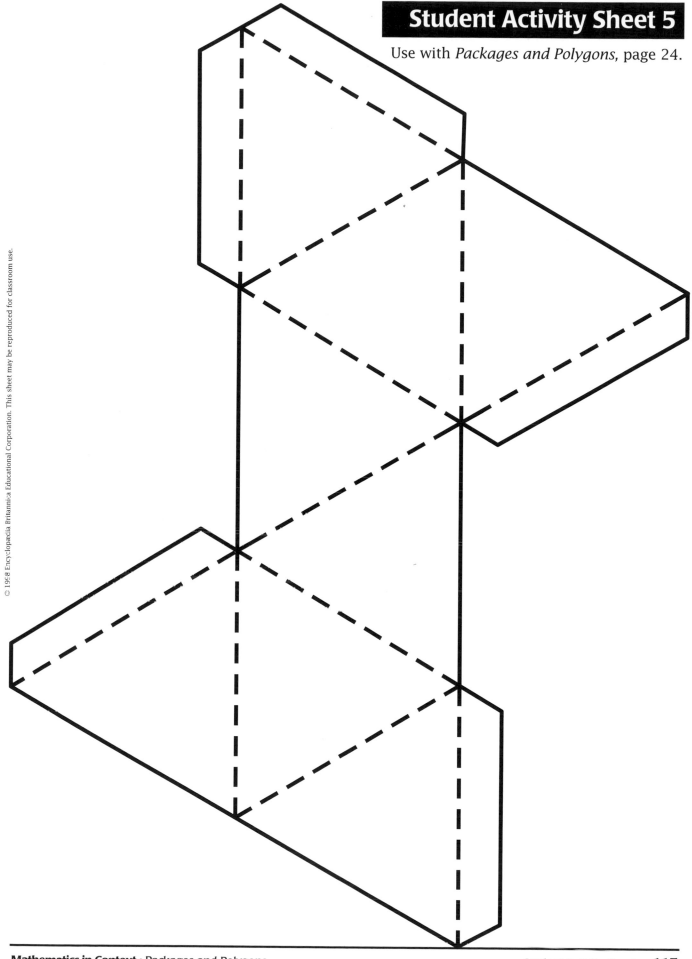

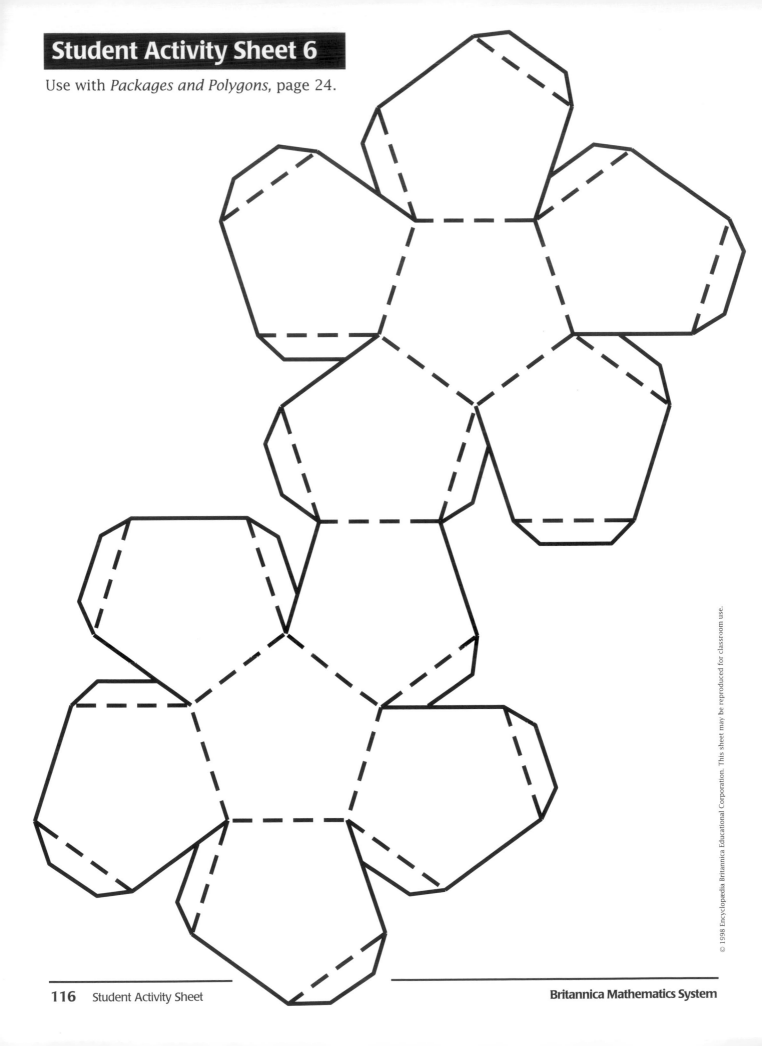

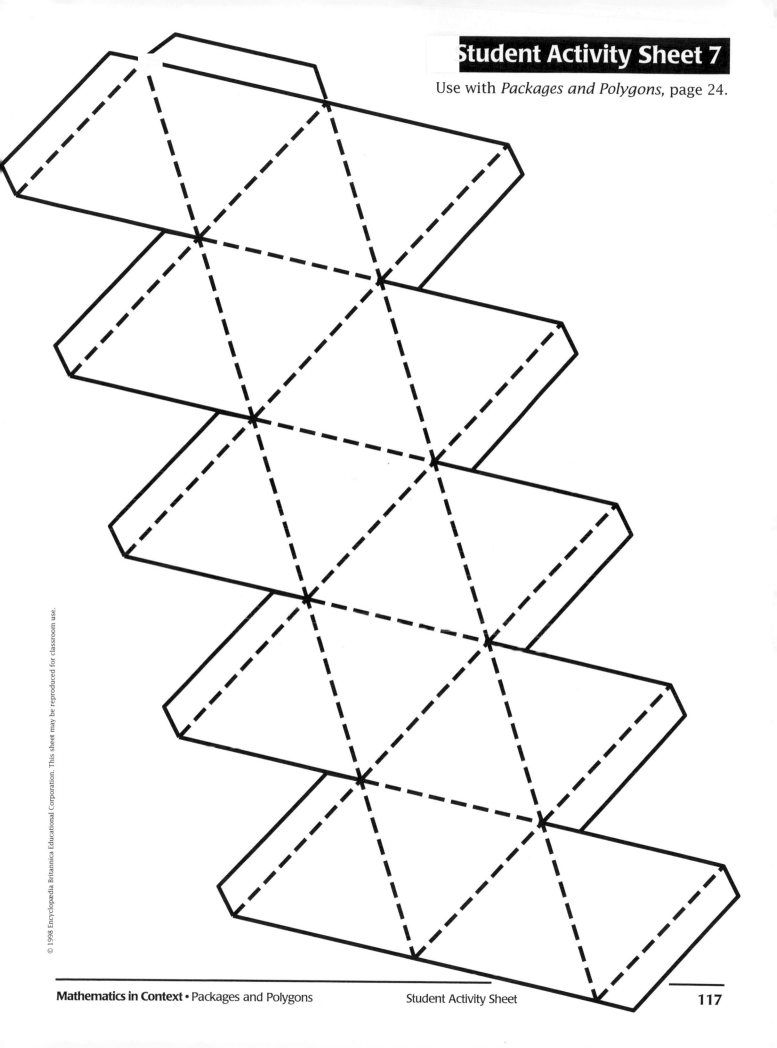

9. Color one face of the tetrahedron in each of the pictures below.

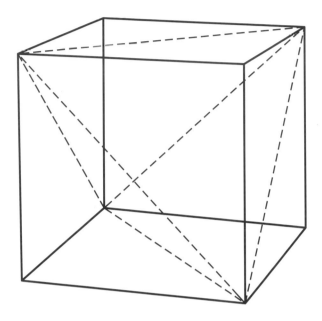

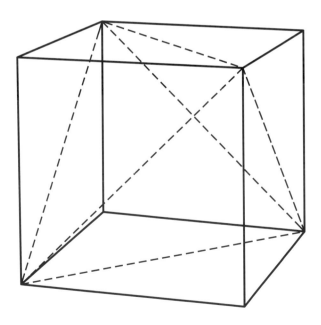

Name_____

11. a. Complete the following table. To help you, use the five models you made.

b. Study the numbers in your table. What patterns or relationships do you see?

Name	Shape	Type of Face	Number of Faces	Number of Vertices	Number of Edges
Tetrahedron		triangle	4	4	6
Cube					
Octahedron					
Dodecahedron					
Icosahedron					

Use with
Packages and Polygons,
page 33.

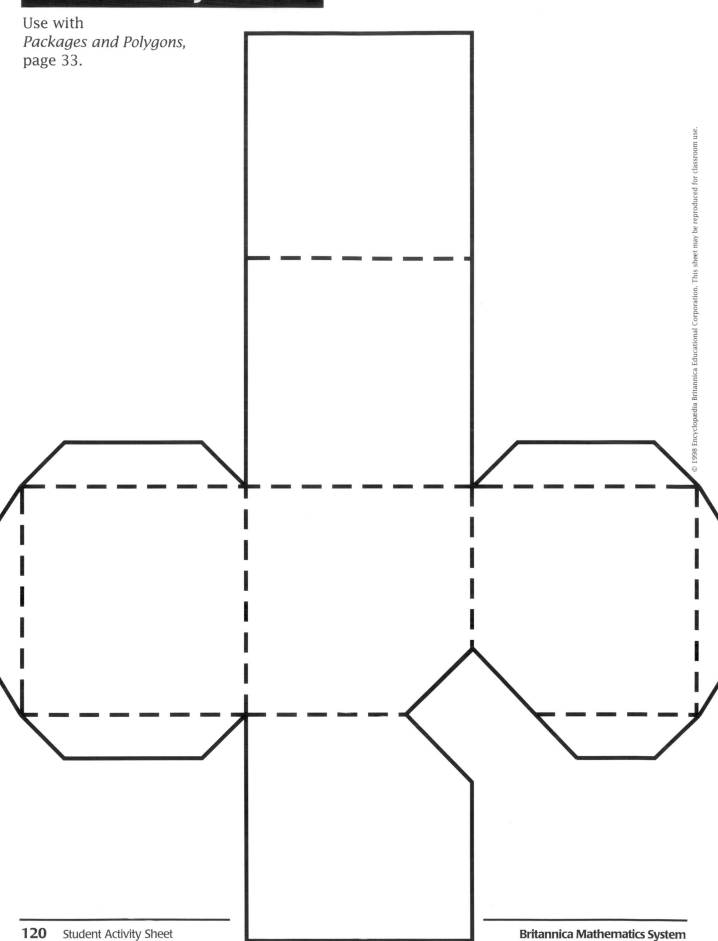

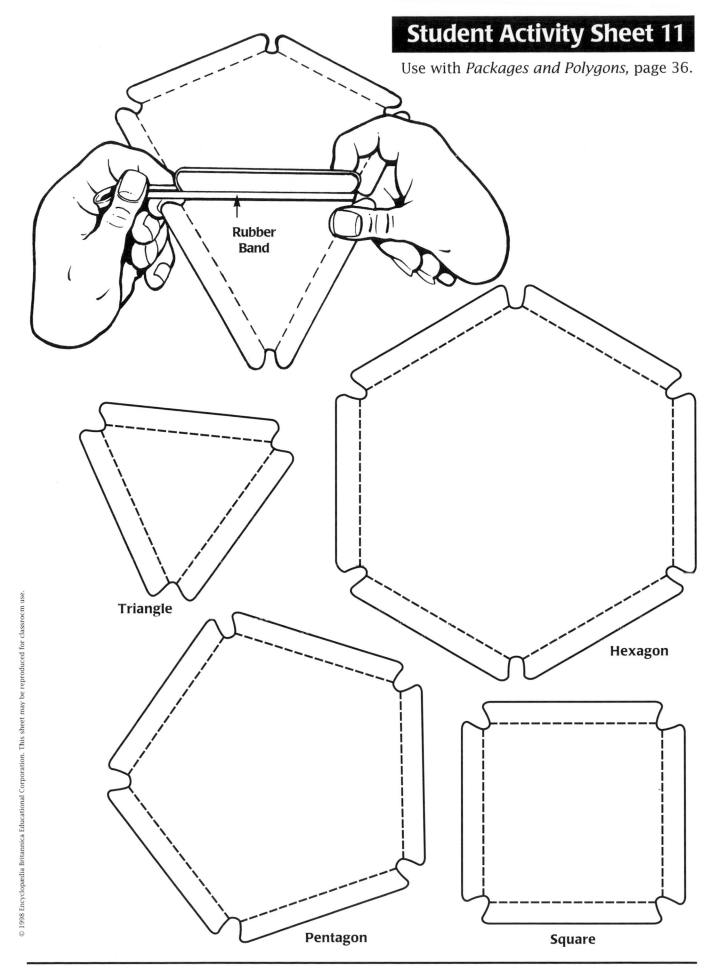

Rubber Band

Triangle

Hexagon

Pentagon

Square

Name _____

Use with *Packages and Polygons,* page 36.

5. Study your semi-regular polyhedra carefully to find *F*, *V*, and *E*. Check your results using Euler's formula and fill in the following table.

Shape	Types of Faces	F	V	E	F+V−E
	Triangle and Hexagon	14		18	
		60			

7. a. Draw the glue strips that are necessary to build the boxes from these nets.

b. Trace your patterns onto heavy paper and build the boxes.

NAMING SHAPES

Use additional paper as needed.

1. Name the three-dimensional shapes you recognize in the objects pictured below.

a.

b.

c.

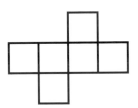

d.

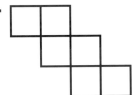

2. How would you explain to someone how to tell the difference between a pyramid and a prism?

3. Is it possible to make a cube from any of the nets pictured below?

a.

b.

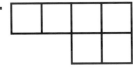

c.

4. The cube below is painted half gray and half white. Draw two different nets for this cube. Be sure to color the faces.

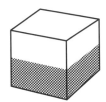

Britannica Mathematics System

POLYHEDRA CONSTRUCTIONS

Use additional paper as needed.

1. **a.** The structure on the right can be made of wood. What shapes do you have to glue together to make this structure? How many do you need?

   b. With what shape could you fill the corners to make the structure into a regular polyhedron?

   c. If you fill all the corners, what shape will the whole structure have?

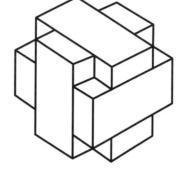

2. The drawings on the right are of models of polyhedra.

   a. Figure i is built from two tetrahedra. Color one tetrahedron red and the other one blue.

i.

   b. From which two polyhedra is figure ii constructed? Color each one in a different color.

ii.

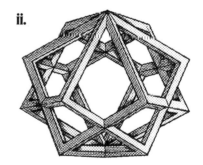

   c. Figure iii is constructed from three regular polyhedra. Color one of these. What is the name of this polyhedron?

iii.

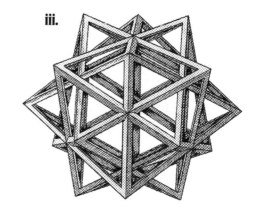

TILING WITH POLYGONS

Use additional paper as needed.

1. This is a tiling composed of regular polygons.

    a. Name the regular polygons you recognize.

    b. Look at the point labeled 1. You can find four angles with a vertex at 1. What is the measure in degrees of each of these angles? Explain how you found your answer.

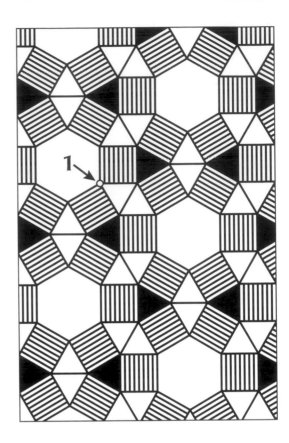

2. One tip is cut off of a cube. On the opposite vertex of the cube, the one that is hidden in the picture on the right, a similar piece will be cut off.

    a. How many vertices will the new shape have?

    b. How many faces?

    c. How many edges?

    d. In the space on the right, write a formula for the relationship between the numbers of faces, vertices, and edges using *F*, *V*, and *E*. Use your formula to check your answers to parts **a, b,** and **c.**

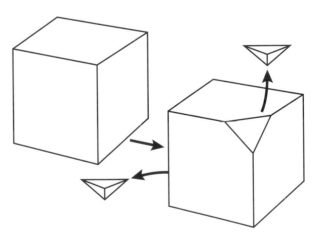

HIDDEN FACES

Use additional paper as needed.

Pictured here is a view of a solid. The hidden half is identical to the visible half.

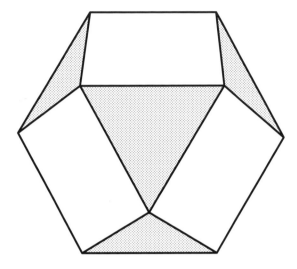

1. How can you make this shape from a cube?

2. Suppose you were to make a bar model of this shape. How many edges do you need? How many vertices do you need? Explain how you found your answers.

ALTERNATIVE ASSESSMENT

Use additional paper as needed.

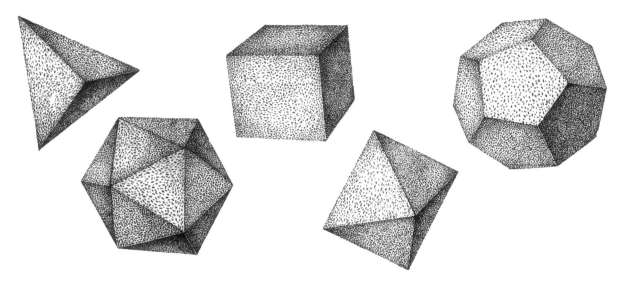

1. Your teacher has placed a number of solids at the front of the room. Your group will choose one or two of these solids to make a design. First, your group will decide what kind of object you are going to make. You may decide to design a package, an art object, an object that could be used at a playground, or something else. Next, choose one or two of these solids and make your design; you can paste solids together or cut off pieces so you will get semi-regular polyhedra.

2. Write a report about your design. Be sure that your report includes the following information:

   • the purpose of your design,

   • detailed information about the shape, and whether or not Euler's formula is true for it,

   • directions for making the design, including a net.

3. Present your design to the class.

Section A. Packages

1. Answers will vary. Sample responses:

    a. baseball

    b. shoe box

    c. funnel

    d. coffee can

2. **a.** cube

    b. rectangular prism

    c. pyramid

    d. cylinder

    e. prism

3–4. a. Rectangular Prism

        b. Rectangular Prism

        c. Truncated Pyramid

Rectangular Prism

        Rectangular Prism

        Pyramid

   d.   Cylinder

        e.   Prism

Cylinder

        Prism

5. The new shapes in problems **3** and **4,** parts **b, d,** and **e** have the same names as the original shapes from which they were cut. One of the new shapes in part **c,** the pyramid, has the same name as the original shape from which it was cut.

Section B. Nets

1. Only the net in part **a** can be folded into a rectangular prism.

2. The net in problem **1,** part **b** cannot be folded into a rectangular prism because it is missing one of its square bases. The net in problem **1,** part **c** cannot be folded into a rectangular prism because the two square bases are located on the same side of the net.

3. No. The two cones do not have the same height because they have different-sized bases. Compare the drawings on the right:

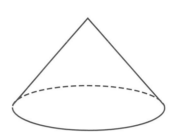

Section C. Bar Models

1. **a.** Peter needs a total of 12 bars to build a bar model of a rectangular prism.

   b. Of the 12 bars, Peter needs four bars that are each 7 centimeters long, four bars that are each 3 centimeters long, and four bars that are each 5 centimeters long. See the drawing on the right.

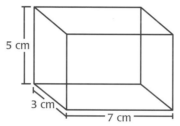

2. **a.** Kim needs eight pieces of wire (or eight edges) to build the pyramid.

   b. Answers will vary. Kim wants the base of the pyramid to be a square. That means the four pieces of wire forming the base must all have the same length. She also wants to use all 100 centimeters of wire.

   So, one possibility is four pieces of wire that are each 10 centimeters long and four pieces of wire that are each 15 centimeters long; 4 × 10 cm = 40 cm and 4 × 15 cm = 60 cm for a total of 40 cm + 60 cm = 100 cm, the total length of the wire.

   Another possibility is four pieces of wire that are each 5 centimeters long and four pieces of wire that are each 20 centimeters long; 4 × 5 cm = 20 cm and 4 × 20 cm = 80 cm for a total of 20 cm + 80 cm = 100 cm, the total length of wire.

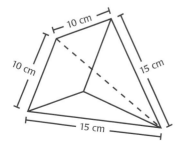
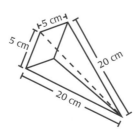

3. a. Four faces are hidden.

b. Five faces have the shape of a rectangle.

c. Five edges are hidden.

d. Two vertices are hidden.

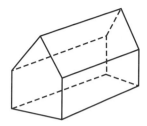

Section D. Polygons

1. a.

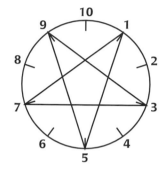

b. Inside the star is a 5-gon, also known as a pentagon:

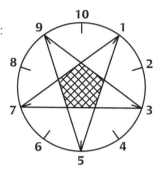

c. 36°. Explanations may vary. Sample explanation:

I know that each turn needed to make a pentagon is 72° (360° ÷ 5); this is the same as the measure of each angle in the triangles that is adjacent to the pentagon. Since the angles of a triangle total 180°, and 72° + 72° = 144°, the smallest angle in each triangle (the star point) measures 180° − 144°, which is 36°.

2. A 5-gon (pentagon) and a 10-gon can be made with equal-sized jumps from one whole number to another on the diagram.

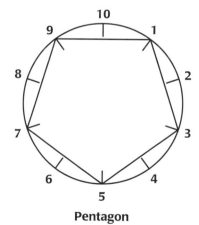

Pentagon

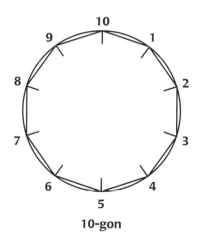

10-gon

3. Yes. Students may reason that there are 10 whole numbers on the diagram. The factors of 10 are 10, 5, 2, and 1. So, the polygons that can be formed by making equal-size jumps from one whole number to another on the diagram should be factors of 10. As shown in problem **2,** a 10-gon and a 5-gon were the only polygons that could be made in this manner. (*Note:* There is no such thing as a 2-gon or a 1-gon because polygons must have three or more sides or angles.)

Section E. Platonic Solids

1. **a.** The tetrahedron, octahedron, and icosahedron are polyhedra with faces in the shape of a regular triangle. These are regular polyhedra, also known as Platonic solids.

   There are also three semi-regular polyhedra with one or more faces in the shape of a regular triangle. These semi-regular polyhedra are formed by cutting the corners off the following Platonic solids: a tetrahedron, a cube, and a dodecahedron.

   b. There are three semi-regular polyhedra with one or more faces in the shape of a hexagon. These semi-regular polyhedra are formed by cutting the corners off the following Platonic solids: a tetrahedron, an octahedron, and an icosahedron.

2. **a.** No. Since polygons share edges and vertices, some of the octahedron's eight edges that are visible in the picture are shared by the side of the octahedron that is not visible.

   b. Four of the edges that are visible in the picture are shared by the back of the octahedron, and four of the visible edges are not shared. That makes a total of eight visible edges and four hidden edges for a total of 12 edges.

3. Toni can make two octahedrons from the bar model of one icosahedron. One icosahedron has 12 vertices and 30 bars. An octahedron requires six vertices and 12 bars, so two octahedrons require twice as much, or 12 vertices and 24 bars. Three octahedrons require more vertices and bars than are found in one icosahedron.

Section F. Euler's Formula

1. Yes. There are seven faces, 10 vertices, and 15 edges on the following polyhedron. Euler's formula states that $F + V - E = 2$. Here, $7 + 10 - 15 = 2$, so Euler's formula works for this polyhedron.

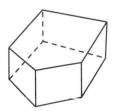

Section G. Semi-regular Polyhedra

1. Explanations may vary. Students should note that Euler's formula states that $F + V - E = 2$ for all polyhedra. So, if it is possible to build a polyhedron with nine vertices and five edges, then $F + 9 - 5$ would have to equal 2. However, F would have to be -2 in order to solve the equation. Since every polyhedron has at least one face, this structure is not possible.

Cover

Design by Ralph Paquet/Encyclopædia Britannica Educational Corporation.

Collage by Koorosh Jamalpur/KJ Graphics.

Title Page

Paul Tucker/Encyclopædia Britannica Educational Corporation.

Illustrations

6 Paul Tucker/Encyclopædia Britannica Educational Corporation; **12 (e, h, j)** Phil Geib/Encyclopædia Britannica Educational Corporation; **12 (c, g, i)** Paul Tucker/Encyclopædia Britannica Educational Corporation; **22, 24, 26** Phil Geib/Encyclopædia Britannica Educational Corporation; **30, 32** Paul Tucker/Encyclopædia Britannica Educational Corporation; **34, 36, 60, 62, 64, 82, 84, 86** Phil Geib/Encyclopædia Britannica Educational Corporation; **88** Paul Tucker/Encyclopædia Britannica Educational Corporation; **90** Phil Geib/Encyclopædia Britannica Educational Corporation; **98** Paul Tucker/Encyclopædia Britannica Educational Corporation.

Photographs

34 © Robert Drea; **46** Department of Defense/Hammond; **52** © Grant Heilman/Grant Heilman Photography, Inc.; **62** Photograph Courtesy of GE Superabrasives; **72** © North Wind Picture Archives; **84** © Robert Drea.